Computer Simulation S
Phenomena in Dense Su
Blood Cells under Shea

Timm Krüger

Computer Simulation Study of Collective Phenomena in Dense Suspensions of Red Blood Cells under Shear

Timm Krüger
Düsseldorf, Germany

ISBN 978-3-8348-2375-5 ISBN 978-3-8348-2376-2 (eBook)
DOI 10.1007/978-3-8348-2376-2

The Deutsche Nationalbibliothek lists this publication in the Deutsche Nationalbibliografie; detailed bibliographic data are available in the Internet at http://dnb.d-nb.de.

Springer Spektrum

Cover design: KünkelLopka GmbH, Heidelberg

Printed on acid-free paper

Springer Spektrum is a brand of Springer DE. Springer DE is part of Springer Science+Business Media.
www.springer-spektrum.de

Blut ist ein ganz besondrer Saft.

Johann Wolfgang von Goethe (Faust)

Abstract

Understanding the rheology of blood has been a scientific task for nearly one century—for good reasons. On the one hand, blood is vital for the human body, and some diseases like malaria or sickle-cell anemia interfere with its proper functioning. On the other hand, blood is an example for a dense suspension. Up to now, the rheological and dynamical properties of such complex fluids is not completely understood. For this reason, investigating the properties of dense suspensions in general and those of deformable particle suspensions in particular is of paramount importance. Due to the highly complex boundary conditions which can be found in suspensions on the particle scale, analytic investigations alone cannot clarify the unanswered questions. Especially in the first decade of this century, the computational resources and available algorithms have become mature enough to allow numerical studies of suspensions of deformable particles. This is the primary aim of the current work.

In the present thesis, a numerical tool is developed which allows to simulate particulate blood suspensions and to investigate their mechanical properties. Due to the separation of cellular and molecular time and length scales, the basic idea is to follow a mesoscopic approach. The red blood cells are resolved, and their deformation state is tracked explicitly. However, the suspending fluid is described as a continuum medium with Newtonian properties. The lattice Boltzmann method (chapter 5) is employed as Navier-Stokes solver for the suspending fluid, whereas a finite element method (chapter 7) is used for the description of the elastic red blood cell membranes. A fluid-structure interaction algorithm is required to couple the particle motion and the fluid dynamics. The immersed boundary method (chapter 6) serves this purpose well. The model is extended in such a way that dense suspensions (up to 65% volume fraction) of $\mathcal{O}(1000)$ deformable particles in arbitrary geometries can be simulated with reasonable computational effort (chapter 8).

During the course of this thesis, a new boundary condition for the lattice Boltzmann method is introduced (section 5.4). This extension of the bounce-back boundary condition can be used to impose a well-defined shear stress to drive the fluid, even when its viscosity is not known. This is particularly important when the static yield stress of a suspension is to be investigated. The static yield stress is defined as the stress below which no flow occurs. Contrarily, shearing a suspension with a finite shear rate does not allow to find the static yield stress. Instead, the dynamic yield stress (as the stress for vanishing small shear rates) may be obtained which is usually smaller than the static yield stress.

In appendix B, it is argued and shown via simulations that the shear stress in the lattice Boltzmann method is a second-order accurate observable. On the one hand, the focus of such investigations is usually on the velocity as it is the most relevant observable for many hydrodynamic problems. For suspension rheology, on the other hand, the stress plays a more important role. A consistent picture of stress evaluation in immersed boundary lattice Boltzmann simulations is provided in chapter 9. It is shown how the particle contribution to the suspension stress can be computed locally (in space and time) and independently from the fluid stress or macroscopic assumptions. This approach is especially important when spatio-temporal fluctuations of the shear stress are sought after. These fluctuations carry significant information about the statistical properties of the suspension.

The computational model is utilized to study the rheology of blood and its microdynamics systematically (chapter 10). The influence of the most important control parameters (the shear rate $\dot{\gamma}$, the volume fraction Ht, and the red blood cell rigidity κ_S) is investigated. It is found that the model recovers the experimentally obtained flow curve for blood at intermediate volume fractions. The particle deformability significantly affects the microdynamics in the suspension: When the suspension stress exceeds a certain threshold, the red blood cells start to tank-tread, and an increased orientational ordering develops. The combined effect of tank-treading and collective alignment is one of the mechanisms contributing to the shear thinning behavior of blood. The simulations provide clear evidence for the importance of a correct microscopic description of the red blood cells in simulations. Although suspensions of rigid particles also exhibit shear thinning under some circumstances, the microscopic behavior of the suspensions is found to be significantly different. A remarkable result related to the simulation parameter space is found. All relevant results can be described by two, rather than three parameters, the volume fraction Ht and the capillary number Ca $\propto \dot{\gamma}/\kappa_S$. Thus, the effect of varying particle rigidities can be compensated by the shear rate. An interesting result is that some of the data (e.g., the particle tumbling frequency, the deformation, or the collective order parameter) collapse on a *single* master curve when plotted as function of the 'corrected' capillary number $\mathrm{Ca}^* \propto \sigma/\kappa_S$ (σ is the total suspension stress). These observations can be explained by the idea that, independent of the suspension volume fraction, tank-treading red blood cells are aware of their neighborhood only via the suspension stress σ.

For the first time, the shear-induced diffusion of red blood cells in simple shear flow is investigated as function of the above-mentioned control parameters ($\dot{\gamma}$, Ht, κ_S). It is shown that the deformability increases the shear-induced Péclet number as compared to a system of rigid particles with comparable volume fraction (section 10.7). Consequently, diffusive motion sets in later and is less efficient in mixing the suspension. It is also found that the fluctuations of the shear rate and the particle shear stress are correlated (section 10.8). The higher the shear rate, the smaller the stress and vice versa. This observation may point at one possible mechanism for stress relaxation which is not relevant for frictionless hard sphere systems: When particles are locked during shearing, they cannot rotate and decelerate the ambient fluid. The stress increases. At some point, the stress is sufficiently large for the particles to be freed again. They increase their angular velocity, the shear rate grows as well, and the stress may relax.

Contents

List of Figures

List of Tables

1. Motivation, aims, and outline of the thesis

1.1. Motivation

Both for medical sciences and fundamental research, blood is an interesting and important suspension. The function of blood is to supply the cells in the human body with oxygen and nutrients and transport waste products away from the cells. The deformability of red blood cells and the fact that human blood is a dense suspension (45% volume fraction) are primary motivations for a better understanding of the rheology of suspensions of deformable particles in general. Even without considering the biochemical and biophysical properties of red blood cells, their mechanical features alone (e.g., deformability and non-trivial shape) give rise to complex rheological features.

Even the dynamics of an isolated red blood cell is still a matter of debate. For example, due to their deformability, red blood cells show varying behavior when the ambient shear rate is changed. Increasing the shear rate, red blood cells are progressively deformed until they 'tank-tread', i.e., the membrane rotates about its own perimeter. This peculiar feature, in connection with the collective cell alignment, is known to be one of the key ingredients of the shear thinning of blood. On this account, it is not a surprise that the collective dynamics of red blood cells in a dense system is not well understood.

Certainly, blood is one of the paramount examples of complex fluids. Yet, it is only one among myriads of others. Suspensions and emulsions are important substances for the industry and everyday life (e.g., food production, pharmaceutics, oil industry, slurry handling, to name only a few). Until today, it is still not entirely clear which microscopic properties of a complex fluid determine its macroscopic rheology. One of the ultimate goals is to identify mechanisms common for wide classes of complex fluids, independent of the details of the system. Which macroscopic features of a complex fluid are general, and which are specific to a given subclass? How can the fluid rheology be predicted based on its microscopic features? It is reasonable to better understand the rheology of blood as a specific example before it is compared to other complex fluids. In fact, it is expected that the insights gained from these investigations can be applied to the study of other complex systems too.

The analytical solution of the microscopic equations of motion for a complex fluid remains a technical challenge, in particular due to the complex boundary conditions at the particle surfaces and the required description of fluid-structure interactions. In order to investigate, via numerical methods, the macroscopic rheological behavior based on the microscale physics, it is important to simulate a large number of particles providing a meaningful statistical description. At the same time, the mechanical and dynamical behavior of individual particles should be captured with sufficient accuracy. This way, events on the micro- and the macroscales can be correlated, resulting in a better understanding of the scale-bridging mechanisms.

The above considerations call for efficient and accurate numerical methods capable of handling complex boundary conditions and hundreds or thousands of resolved deformable particles. One of the available simulation tools in this context is the relatively new and successful lattice Boltzmann method as computational fluid dynamics solver in combination with the immersed boundary method for the fluid-structure interaction.

1.2. Aims of the present work

Due to the complexity of boundary conditions in dense suspensions, computer simulations are indispensable for the study of these systems. Therefore, a large part of the current work has to be dedicated to the development and implementation of a simulation tool for the modeling of dense suspensions of deformable particles in externally applied flows. This tool shall allow to track both the individual behavior of the suspended particles and the macroscopic properties of the suspension at the same time. On the one hand, the particles have to be designed as resolved objects whose shape and deformability can be directly controlled by the user. On the other hand, the numerical model must be sufficiently efficient to allow the simulation of hundreds of particles in order to gain inside into collective effects and macroscopic rheology. This technical challenge requires a thorough debate about the available state of the art in computational fluid dynamics and fluid-structure interaction.

The next step is to provide a closed picture how to control the simulation parameters and to access and evaluate the relevant observables of the simulations. These, in particular, include the global and local fluid and particle stresses and the individual and collective particle dynamics, such as their deformation, rotation, and orientational alignment. Computer simulations are useless if their outcome cannot be analyzed and compared to theoretical expectations, experimental measurements, or results of other computer simulations.

The final goal of this thesis is to extend the knowledge and understanding of the microscopic mechanisms responsible for the shear thinning behavior of blood. Especially the deformation and the tank-treading of the blood cells are inspected. To this end, the results obtained from the developed simulation tool are evaluated and analyzed. Yet, the simulation algorithms and the evaluation methods for the microscopic and rheological observables shall not be limited to the simulation of blood alone. Rather, the methods developed and described in the present work shall be, at least partially, applicable to similar systems as well, such as dense suspensions of arbitrary deformable capsules or vesicles. The focus of the work is on the connection between the collective properties of the system and the dynamics of individual particles. Biological or chemical effects are not part of the investigations, i.e., the analysis relies on the dynamical and mechanical properties of the suspensions and their constituents alone.

1.3. Thesis outline

The thesis is divided into four major parts.

The first part is dedicated to the introductory chapters. A brief overview of complex fluids and their rheology (chap. 2) is followed by an outline of the physics of red blood cells and hemorheology (chap. 3).

In the second part, the selection, setup, and implementation of the numerical model is discussed. After the major physical ingredients of the model have been identified (chap. 4), the lattice Boltzmann method as Navier-Stokes solver (chap. 5), the immersed boundary method as fluid-structure interaction handler (chap. 6), and the membrane model (chap. 7) are presented. Additionally, advanced model considerations, such as mesh generation, benchmark tests, and unit conversions are discussed (chap. 8).

The microscopic and macroscopic stress evaluation is thoroughly investigated and tested in chap. 9. The red blood cell suspension simulations and their detailed interpretations are given in chap. 10, followed by the conclusions and the outlook in chap. 11.

The appendix as fourth major part is intended for providing a reference for the conventions used in the thesis (chap. A), some tedious and technical calculations concerning the lattice Boltzmann method (chap. B), and the red blood cell membrane energetics (chap. C).

Part I.

Introduction

2. Introduction to complex fluids and their rheology

This chapter is intended to provide a brief overview of rheology in general (section 2.1), the definition of complex fluids and the challenge of their investigation (section 2.2), and the rheology of suspensions made of rigid particles (section 2.3) and deformable objects (section 2.4). A review of the physics of red blood cells and hemorheology is given in chap. 3.

2.1. Introduction to rheology

Rheology (from Greek ρει 'to flow' and Greek λογοζ 'reason') is the study of the flow and deformation of matter in general and fluids in particular.

The macroscopic equations of motion for an arbitrary, incompressible fluid read

$$\rho\left(\frac{\partial \boldsymbol{u}}{\partial t}+(\boldsymbol{u}\cdot\nabla)\boldsymbol{u}\right)=-\nabla p+\nabla\cdot\boldsymbol{\sigma}+\boldsymbol{f} \tag{2.1}$$

and

$$\nabla\cdot\boldsymbol{u}=0. \tag{2.2}$$

Here, ρ, p, and $\boldsymbol{u}$ are the density, isotropic pressure, and the velocity of the fluid. The deviatoric stress tensor is denoted by $\boldsymbol{\sigma}$, and external force densities (e.g., gravity) by $\boldsymbol{f}$. Eq. (2.1) reflects Newton's second law, and eq. (2.2) is the continuity equation for the incompressible fluid. These equations are only valid for continuum fluids, i.e., the individual motion of atoms and molecules are not resolved. At length scales of about 100 molecular radii, one can consider fluids a continuum [1].

The momentum flux tensor is defined as

$$\boldsymbol{M}:=\rho\boldsymbol{u}\boldsymbol{u}+p\boldsymbol{I}-\boldsymbol{\sigma} \tag{2.3}$$

where $\boldsymbol{I}$ is the identity matrix. This way, the components of eq. (2.1) can be written in the compact form $\rho\partial_t u_\alpha=-\partial_\beta M_{\alpha\beta}+f_\alpha$. Often, the pressure p and the deviatoric stress tensor $\boldsymbol{\sigma}$ are combined to the total stress tensor $\boldsymbol{\Sigma}=-p\boldsymbol{I}+\boldsymbol{\sigma}$. This stress tensor describes the part of momentum flux not being related to the mass transfer. The total stress tensor $\boldsymbol{\Sigma}$ corresponds to the force $\mathrm{d}\boldsymbol{F}$ exerted by the fluid on an oriented area element $\mathrm{d}\boldsymbol{A}$ such that $\mathrm{d}\boldsymbol{F}=\boldsymbol{\Sigma}\cdot\mathrm{d}\boldsymbol{A}$.

The deviatoric stress tensor $\boldsymbol{\sigma}$ in eq. (2.1) contains all stress contributions which are anisotropic. The deviatoric stress is an important quantity as it is a measure for the dissipation of energy in the fluid [2]. In order to solve eq. (2.1), $\boldsymbol{\sigma}$ has to be known as function of the fluid velocity $\boldsymbol{u}$ and its derivatives. The relation $\boldsymbol{\sigma}(u_\alpha,\partial_\alpha u_\beta,\ldots)$ is called the constitutive law of the fluid. One can show that, for not too large velocity gradients, the stress tensor is only a function of the first derivatives of the velocity, $\boldsymbol{\sigma}=\boldsymbol{\sigma}(\partial_\alpha u_\beta)$ [2]. For a Newtonian fluid, the constitutive law is

$$\boldsymbol{\sigma}=2\eta_0\boldsymbol{S} \tag{2.4}$$

where η_0 is a constant and

$$S_{\alpha\beta} := \frac{1}{2}\left(\partial_\alpha u_\beta + \partial_\beta u_\alpha\right) \tag{2.5}$$

is the symmetrized velocity gradient tensor. In simple shear flow, the tensor $\boldsymbol{S}$ has only one non-trivial off-diagonal component, the shear rate $\dot{\gamma}$. The corresponding component of the deviatoric stress tensor $\boldsymbol{\sigma}$ is the so-called shear stress. The proportionality constant η_0 is also denoted the dynamic shear viscosity of the fluid. It is a macroscopic material property and a measure for the resistance to flow under applied stresses. The viscosity can be experimentally obtained with viscometers and rheometers. These devices rely on the possibility to measure shear rate and shear stress independently [3].

It has to be emphasized again that eq. (2.1), the concept of the deviatoric stress $\boldsymbol{\sigma}$, and the viscosity η_0 are only reasonable on the macroscale where the fluid can be considered a continuum. If the fluid is described on the molecular level by solving Newton's equations of motion for each molecule, it is not directly clear how the stress and viscosity shall be defined. In particular, the macroscopic stress tensor derives from the microscopic interactions and forces between the molecules, and it is difficult to obtain local expressions for the stress tensor on the microscopic scale where interactions are usually non-local [4, 5]. Within the framework of statistical physics, it is possible to give expressions for the macroscopic stress and viscosity for simple systems (e.g., an ideal gas) as function of the microscopic properties of the system [6]. Additionally, it has to be noted that only the divergence of the stress tensor enters the equations of motion, eq. (2.1), which leaves a gauge freedom in the stress definition.

Many ubiquitous 'complex fluids' (section 2.2) like blood, honey, or mayonnaise are non-Newtonian, i.e., their flow curves (viscosity versus shear rate) are not constant. If the viscosity decreases with increasing shear rate, the fluid is called shear thinning. In the opposite case, it is denoted shear thickening [1, 7]. One of the problems when studying these fluids is that the constitutive laws $\boldsymbol{\sigma}(\boldsymbol{S})$ are only valid for certain classes of fluids. The detailed constitutive law found empirically for one material usually fails to describe another. For each fluid, the constitutive law has to be measured experimentally. These results can be compared with theoretical predictions based on statistical physics and microscopic models for the fluids. In fact, it is one ultimate goal to identify the common and distinct features of wide classes of complex fluids [1, 7, 8]. Some of the mechanisms for the non-Newtonian behavior are presented in sections 2.3 and 2.4.

2.2. The complexity of complex fluids

The term 'complex fluid' denotes substances which flow at modest stresses on time scales perceivable by humans. On the microscale, complex fluids usually have relevant length scales which are much larger than atomic ones. This can lead to large structural relaxation times and non-Newtonian effects, especially for dense systems. Complex fluids are neither simple liquids nor simple crystalline solids, they are 'in between', and the constitutive law is usually non-linear [3]. Examples for complex fluids are mustard, chocolate, toothpaste, asphalt, paints, blood, polymer gels, foams, or liquid crystals, to name only a few. It is obvious that the understanding of complex fluids is of paramount importance for medical and industrial applications, but also for fundamental research.

Suspensions and emulsions are two prominent subclasses of complex fluids. Suspensions are systems of rigid or deformable particles distributed in a simple liquid. Blood as a suspension of deformable red blood cells is one of the most important examples (chap. 3). An emulsion is a system of two immiscible liquid phases. The viscosity ratio of the emulsified and the emulsifying phase is denoted λ. Emulsions are important for the petroleum, food, medical, or cosmetics industries [9]. When dealing with dense systems, 'dense' means that the average distance of the particles is smaller than the particle size or that the system shows non-Newtonian rheology [1].

For an observer, suspensions and emulsions are only determinate in a statistical sense because it is usually impossible to describe the microstate of each suspended particle in detail [10]. Still, the microstructure has to be studied in order to understand the rheology and the origins of the constitutive behavior of these complex fluids [1]. It is not a surprise that theories for complex fluids become more and more complicated when the particles have non-trivial features such as non-spherical shapes or intrinsic deformability [11]. The basic difficulty of understanding the macroscopic behavior of homogeneous suspensions has been summarized by Batchelor [10]:

> The problem is to determine the rheological properties of this equivalent homogeneous fluid from a knowledge of the properties of the particles and the ambient fluid in which they are suspended; in other words, to determine the relation between the macroscopic or bulk properties of the suspension and its microscopic structure on the particle scale.

This statement can be extended to complex fluids in general, including blood rheology on the macroscale and the properties of red blood cells on the microscale. Batchelor [10] also emphasized that

> a major difficulty in the study of rheology is that one's intuition about the form of the constitutive stress relation appropriate to given circumstances is so poorly developed.

As already mentioned in section 2.1, it is of practical importance to find reliable constitutive laws for complex fluids derived from microscopic properties (e.g., particle shape, interaction, and deformability) and individual particle motion (e.g., particle rearrangements and stress relaxation). General constitutive stress-strain relations are not known for dense complex fluids today, and further research is necessary [1].

For suspensions, there is a clear separation of length and time scales between the suspending medium and the suspended particles. Thus, it is reasonable to use mesoscopic approaches where the medium is considered a continuum and the suspended particles are resolved [11]. Still, one practical reason for the intricacy of these systems is the pronounced complexity of the microscopic boundary conditions at the particle-liquid interfaces. Therefore, the equations of motion for realistic systems can only be solved numerically and computer simulations are necessary. They complement the elaborate experiments performed to gain a better understanding of the connections between the micro- and the macroscale in complex fluids.

In sections 2.3 and 2.4, a brief summary of the rheology of complex fluids made of rigid and deformable objects is given. The rheology of blood is described separately in chap. 3. A more detailed overview of complex fluids and simulations and experiments to investigate their behavior can be found in the literature (e.g., [1, 3, 11, 12]).

2.3. Rheology of complex fluids made of rigid particles

The behavior of even the simplest suspensions (non-interacting, monodisperse, spherical, rigid particles in a Newtonian liquid) is so rich that it cannot be summarized exhaustively at this point. Rather, a brief overview will be given. Recent review articles by Stickel and Powell [1] and Brader [11] and monographs by Dhont [13] and Larson [3] provide additional information.

The dimensionless parameters commonly used to characterize suspensions of rigid particles (particle radius r, suspending fluid viscosity η_0, external shear rate $\dot{\gamma}$) are the volume fraction, ϕ, the particle Reynolds number, $\mathrm{Re} := \rho\dot{\gamma}r^2/\eta_0$, the thermal Péclet number $\mathrm{Pe} := \dot{\gamma}r^2/D_{\mathrm{th}}$ (where D_{th} is the thermal diffusivity), and the Weissenberg number $\mathrm{Wi} := \dot{\gamma}\tau$ (where τ is a structural relaxation time) [11]. For dilute systems, the relaxation time obeys $\tau \propto 1/D_{\mathrm{th}}$, and the

Péclet and Weissenberg numbers are equivalent. For dense suspensions, however, the structural relaxation time may be much longer than predicted by the bare particle diffusion.

The presence of particles immersed in the suspending liquid increases its viscosity. Einstein [14] was the first to determine the apparent viscosity of a dilute suspension of rigid, spherical particles in simple shear flow,

$$\eta = \eta_0 \left(1 + \frac{5}{2}\phi + \mathcal{O}(\phi^2) \right) . \tag{2.6}$$

'Dilute' means that effects of order ϕ^2 are small compared to the linear corrections. This is the case for volume fractions below a few percent. Eq. (2.6) reveals that these kind of suspensions are still Newtonian because the leading correction term does not depend on the shear rate.

The viscosity for suspensions of hard spheres can be expanded in terms of ϕ where each power of ϕ corresponds to hydrodynamic interactions of this order. For example, ϕ^2-terms correspond to two-particle interactions. Thus, for high volume fractions, analytic expressions are difficult to obtain [3]. The rheology of suspensions becomes more and more complicated when the volume fraction reaches some maximum value ϕ_{m} where the viscosity diverges (see below) [1, 3]. Things become even more difficult when polydispersity is considered because smaller particles can be packed in the holes between the larger ones [3].

In the following, it is assumed that the volume fraction is not too large ($\phi < 0.50$). The interesting case of even denser systems is briefly discussed at the end of this section.

Batchelor [10] has argued that, for an athermal suspension of rigid particles, the suspension will always be Newtonian except the suspending fluid itself becomes non-Newtonian or the particle inertia becomes important. In reality, for Brownian systems, it is observed that suspensions of rigid particles generally show shear thinning behavior at some intermediate shear rate range if the volume fraction exceeds 30 or 40% [3]. Below this regime, at smaller shear rates, the suspensions are Newtonian. The reason for this shear thinning is caused by a change of the suspension microstructure [13]. Shear thinning is related to a relative decrease of the Brownian contribution to the stress [15]. As a result, interparticle distance distributions are modified, particles can arrange in layers, and collisions become less probable, thus reducing the flow resistance [3]. Therefore, non-Newtonian fluids become anisotropic when they are sheared [8].

If the inverse shear rate is small compared to the structural relaxation time τ of the fluid, i.e., if $\mathrm{Wi} = \dot{\gamma}\tau > 1$, the fluid behavior is generally transient and shear thinning. For small Weissenberg numbers, the system is within the linear response regime, and the viscosity is constant [11]. On the one hand, for Newtonian fluids such as water, the relaxation time is of the order of $10^{-12}\,\mathrm{s}$, which explains why non-Newtonian properties of water in practical applications are unimportant [8]. On the other hand, for complex fluids where the suspended particles are orders of magnitude larger, the time scale related to the change of microstructure can be comparable to the simulation or experimental time [1]. Therefore, non-Newtonian behavior is often related to time-dependent rheology, and the history of shearing plays a role. Fluids with decreasing viscosity as function of time are called thixotropic. Non-Newtonian effects in atomic systems are in principle also observable at very high shear rates ($\dot{\gamma} > 10^{10}\,\mathrm{s}^{-1}$). This, however, is orders of magnitudes higher than what can be realized in today's experiments [11]. If the particles are non-spherical, additional effects such as flow-alignment can support the shear thinning behavior [1].

When the shear rate is increased beyond the shear thinning regime, another Newtonian regime is present. After this, at even larger shear rates, shear thickening in dense suspensions can be observed. This effect is generally related to a decay of layers and particle clustering due to lubrication [3, 11]. Shear thickening is found for a wide range of suspension classes at high shear rates [16].

Interesting additional effects can be observed when the suspension volume fraction ϕ exceeds 0.50. For increasing volume fractions, the first Newtonian regime is progressively shifted towards smaller shear rates until it completely vanishes from the experimental window [11, 17, 18]. When the volume fraction reaches values of about 0.58, the particles are trapped in a cage of neighbors, diffusivity decreases drastically, and the viscosity significantly increases until it diverges at $\phi_m \approx 0.63$ which is the random close packing [1, 11]. Above the density of 0.58, the suspension is said to be in the 'glassy state' if the particles are not forming a crystal lattice structure. The transition from the liquid to the glassy state is called *glass transition*. The mode-coupling theory [19] predicts that the viscosity in the glassy state diverges like $\eta \propto \dot{\gamma}^{-1}$ for $\dot{\gamma} \to 0$. In this case, there would exist a finite *dynamic* yield stress $\sigma_y = \lim_{\dot{\gamma}\to 0} \eta(\dot{\gamma})\dot{\gamma}$. The so-called 'creep test' (applied stress to shear the suspension) can be used to find the *static* yield stress [3] below which the initially resting system is deformed elastically and above which plasticity and viscous dissipation set it. The dynamic and static yield stress are generally not identical [20, 21]. A recent overview of the rheology and dynamics of glasses is provided by Barrat and Lemaître [22].

2.4. Rheology of complex fluids made of deformable objects

The rheology of suspensions of deformable particles (such as vesicles or capsules) and emulsions is less understood and has received less attention, especially for high volume fractions. The rheology of these systems is more complicated to understand because the particle deformability requires to solve also the constitutive equations for the particle shapes. Unlike hard spheres, one can produce emulsions with volume fractions larger than the maximum volume fraction ϕ_m. These emulsions are called highly-concentrated. Here, the droplets have to deform even in the unsheared state, and there exists a yield stress above about 80% volume fraction [9]. It has also been shown that blood flow under shear is still possible when the volume fraction is as high as 95% [23, 24]. In this case, the viscosity is governed by lubrication effects.

The deformability of the suspended or emulsified phase introduces another degree of freedom which may be described by an elastic modulus κ or the surface tension γ. For capsules of radius r, the surface elastic shear modulus has unit $[\kappa] = \mathrm{N\,m^{-1}}$ which is identical to that of the surface tension. Therefore, the capillary number $\mathrm{Ca} := \eta_0 \dot{\gamma} r/\kappa$ (γ instead of κ for emulsions) can be defined. It is a measure for the relative importance of hydrodynamic stresses due to shearing and elastic stresses due to particle deformation.

Suspensions of deformable capsules and emulsions are generally shear thinning at any volume fraction [25], which is in marked contrast to dilute suspensions of rigid particles (section 2.3). For the special case of initially spherical capsules in the dilute limit, the apparent viscosity is [26]

$$\eta_a = \eta_0 \left(1 + \phi \left(\frac{5}{2} - \mathcal{O}(\mathrm{Ca}^2)\right) + \mathcal{O}(\phi^2, \mathrm{Ca}^3)\right) \tag{2.7}$$

and decreases with increasing $\mathrm{Ca} \propto \dot{\gamma}$. In the limit of rigid particles ($\mathrm{Ca} \to 0$), Einstein's viscosity for spheres, eq. (2.6), is recovered. A similar result is obtained for emulsions [27],

$$\eta_a = \eta_0 \left(1 + \phi \left(\frac{5\lambda + 2}{2(\lambda + 1)} - \mathcal{O}(\mathrm{Ca}^2)\right) + \mathcal{O}(\phi^2, \mathrm{Ca}^3)\right). \tag{2.8}$$

For $\lambda \to \infty$, the emulsified phase becomes quasi-rigid, $\mathrm{Ca} \to 0$, and Einstein's viscosity is found again. The mechanism for shear thinning of dilute systems of deformable objects is their shear-induced elongation and alignment with the flow [9].

Although both the capsule suspensions and the droplet emulsions are shear thinning, cf. eq. (2.7) and eq. (2.8), the zero-shear viscosities are not identical because the interface physics is different. Due to the absence of elastic stresses in the droplet interface, the interface can contribute to dissipation even when it is not deformed [28]. It has been found that the volume fraction and the deformability alone are not sufficient to describe the macroscopic viscosity because details of the surface play an important role (e.g., surfactants in emulsions or capsule roughness) [9]. This is particularly the case for dense systems where particles are in direct contact.

3. Introduction to the physics of red blood cells and hemorheology

This chapter provides a concise overview of the physics of red blood cells (RBCs) in section 3.1 and the rheology of blood in section 3.2.

3.1. Introduction to the physics of red blood cells

In 1674, after the light microscope had been invented, RBCs have been observed by Anton van Leeuwenhoek for the first time. The average human RBC—also called erythrocyte—is a biconcave cell of about 8 µm diameter and 2.5 µm thickness. Under physiological conditions, its volume and surface are about 100 µm^3 and 130 µm^2, respectively [29]. The shape of a RBC is shown in fig. 3.1. RBCs are anucleate, highly flexible cells consisting of a membrane and the enclosed cytoplasm, a viscous liquid [30] which can only support an isotropic pressure [31]. The cytoplasm is basically a saturated aqueous hemoglobin solution. Hemoglobin is an oxygen carrier protein and responsible for the red color of blood. The main task of a RBC is to transport oxygen to the tissue in the body. In this process, RBCs have to pass capillaries with diameters smaller than half of the large RBC diameter. During this passage, the cells are strongly deformed and have a large contact area with the capillary surface, which enables efficient oxygen discharge by diffusion. Therefore, the deformability is a vital property of a RBC, and it is of paramount importance that the RBC membrane consists of a highly flexible material [31].

Skalak and Branemark [32] observed that the minimum size of a capillary through which a RBC can move is about 3.7 µm. Smaller capillaries would require stronger deformations leading to a surface increase of more than a few percent and consequent hemolysis (membrane damage or rupture) [33]. During its lifetime of about 120 days, a RBC passes the human circulatory system for nearly half a million times [34]. After this period, the flexibility of the membrane has decreased, and the cell cannot move through capillaries efficiently. Hence, it is separated by the spleen and dismantled for recycling. Skalak and Branemark [32] assumed that the spleen may recognize old RBCs by their inability to enter small capillaries.

The RBC membrane is made up of a lipid bilayer (essentially an incompressible 2D fluid with a thickness of 4 nm) and, at its inside, the cytoskeleton. Both leaflets of the bilayer can relax lateral stresses independently, and the required energy to expand or compress the area of such a leaflet is orders of magnitude larger than bending or shearing it [33, 35]. Thus, the total surface of a RBC is practically constant. The cytoskeleton is an elastic network of polymerized proteins supporting the structural stability of the membrane and giving rise to a finite elastic shear resistance [31]. Evans [36] stated that the finite bending resistance of the membrane is caused by the resultant of parallel surface compression resistances in the lipid bilayer. Although the cytoskeleton has an area dilation modulus, it is much smaller than that of the bilayer. The skeleton does not seem to be relevant for the bending resistance either [33, 35]. Additionally, the volume of a RBC is constant as long as the ion concentration in the ambient fluid is not changed. The reason is the impermeability of the membrane with respect to ions and the osmotic pressure caused by ion concentration gradients across the membrane [31]. Due to its partially fluid nature, the RBC membrane also has viscous properties [34, 37, 38]. The nanomechanics of the skeletal network of RBCs is still poorly understood, i.e., it is unclear how the mechanical macroscale properties of

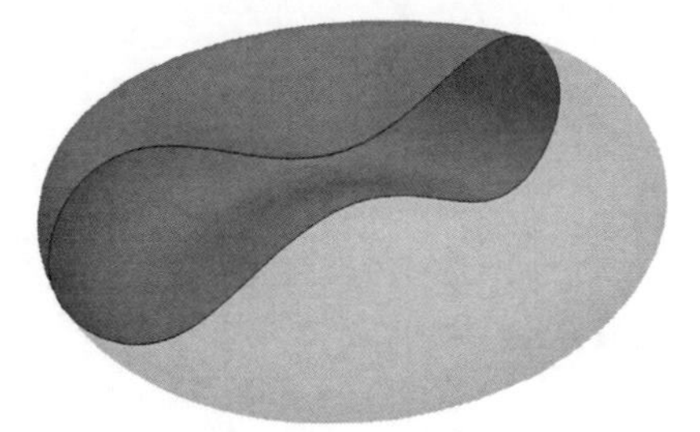

Fig. 3.1.: Schematic shape of a red blood cell (RBC). Half of the cell is made transparent so that the biconcave cross-section (black line) is visible. The in-plane diameter is about 8 µm, the maximum thickness 2.5 µm. The dimples of the RBC are clearly visible.

the cytoskeleton emerge from its microstructure [39].

On a length scale comparable to the RBC diameter, both the lipid bilayer and the cytoskeleton can be regarded as homogeneous and isotropic [31]. Since the membrane surface tension is usually negligible, the dynamics is dominated by curvature and strain elasticity [40]. The bending modulus of the lipid bilayer ($\kappa_B \approx 1.8$–2.1×10^{-19} Nm) has been extracted from different experiments, e.g., micropipette aspiration [41] or local pulling of the membrane with the tip of an atomic force microscope [42]. Finding the modulus for the shear resistance of the cytoskeleton in experiments is rather difficult, and a large scatter of possible parameters have been reported. It is usually assumed that the elastic shear modulus is 'somewhere' in the region between 2 and 10 μN m^{-1} [31, 43]. Generally, the elastic moduli depend on the environment, e.g., pH [44] or osmolarity [45].

The biconcave shape of an undeformed RBC represents the minimization of the membrane energy subject to constant surface and volume constraints. The liquid cytoplasm does not favor any shape, and it can be shown that the cytoskeleton does not play a significant role for the equilibrium energetics either. Thus, the membrane bending energetics determine the shape of a RBC [31].

The small ratio of thickness and lateral extension and the absence of a nucleus make RBCs very flexible [40, 46]. Another ingredient for their large deformability is the surface excess. Compared to a sphere of the same volume, the surface of a RBC is about 25% larger. This additional surface allows the cell to deform significantly, even under the constraint of constant volume and surface. A RBC, therefore, shows similarities with fluid droplets, and blood behaves more like an emulsion than a suspension [30, 47, 48], cf. section 3.2.

Many diseases induce pathological properties of the RBCs, e.g., malaria (an acquired disease) [49], sickle cell anemia (a genetic disorder) [50], or diabetes mellitus [51]. With respect to the mechanical properties of the cells, the above-mentioned diseases are critical because the RBC stiffness is generally increased. As a consequence, it is more difficult for the cells to move through capillaries and to deliver oxygen to the tissue. Moreover, RBCs may die or be damaged prematurely.

3.2. Introduction to hemorheology

Human blood is, by volume, composed of about 55% plasma and 45% suspended blood cells. The volume fraction of the cells is also called hematocrit, Ht. The function of blood is to supply the cells in the human body with oxygen and nutrients and transport waste products away from the cells. Blood plasma is made up mostly of water ($> 90\%$), proteins, glucose, mineral ions, and hormones. It can be considered a Newtonian fluid [7]. The majority of the blood cells are RBCs, cf. section 3.1. Other, less common cells are white blood cells (leukocytes, one for 1000 RBCs [46]) which form a significant part of the immune system and platelets (thrombocytes, one for 15 RBCs [52]) which are relevant for blood clotting. Consequently, blood is basically a dense suspension of deformable RBCs.

Hemorheology denotes the science of the rheology of blood and its particulate components. The largest blood vessels have diameters (a few millimeters up to centimeters) which are more than three orders of magnitude larger than the smallest vessels and the RBCs (a few micrometers). For this reason, hemorheology is a multiscale discipline. Also the range of Reynolds numbers in the human vascular system is huge. On the one hand, in the smallest blood vessels, the flow is basically viscous ($\mathrm{Re} \approx 0.001$ or less). On the other hand, in large arteries, it can reach values of a few 1000 [46, 51]. Understanding hemorheology is relevant mostly for medical reasons. Many diseases are related to modified rheological properties of blood. This can cause further complications, e.g., undersupply of oxygen, stroke, or infarction. Yet, there is growing interest in comprehending the rheology of soft suspensions in general, cf. chap. 2. Review articles about hemorheology can be found in the literature (e.g., [46, 51, 53]).

Blood is a shear thinning fluid for shear rates below $100\,\mathrm{s}^{-1}$. Above, it can be considered Newtonian [7] with a viscosity of 5–6 cP which is about five times larger than that of the suspending plasma [50]. There are different mechanisms for shear thinning in blood: (i) aggregation of RBCs at small shear rates, (ii) deformation, elongation, and tank-treading of RBCs at intermediate and high shear rates [7, 54, 55], and (iii) alignment of RBCs at high shear rates [46]. In general, blood viscosity depends on hematocrit, temperature, plasma viscosity, disease state, age of RBCs, etc. [46].

Schmid-Schönbein and Wells [47] have investigated the shear thinning properties of blood at various hematocrit values. They found that the apparent viscosity of blood at shear rates beyond $230\,\mathrm{s}^{-1}$ is virtually independent of hematocrit (up to more than 90%). This is a hint that RBCs behave similarly to an emulsion under some circumstances, which has also been noted by other researchers (e.g., [30, 48]).

RBCs have been observed to form aggregates at small shear rates (below a few s^{-1}). They can be 1D (rouleaux) or 3D clusters [51, 56, 57]. The protein fibrinogen seems to be responsible for aggregation as it can bind to RBCs and connect them [7, 58]. When the ambient shear rate is increased, the fibrinogen bonds are broken and the aggregates decay. Aggregation and aggregate decay are reversible processes. In the absence of fibrinogen, RBC suspensions behave as a Newtonian fluid at small shear rates [46, 54]. It is believed that the conditions for the buildup of in vivo aggregates are only given in some few veins and venules [59]. The above considerations suggest that RBC aggregation, for physiological hematocrit values, is a pure biochemical mechanism which plays a role only at small shear rates. Blood shows also viscoelastic properties. It is believed that the aggregates rather than individual cells play the dominating role in storing elastic energy at small shear rates [60].

The deformation of RBCs is a key ingredient for its shear thinning behavior for shear rates between a few s^{-1} and about $100\,\mathrm{s}^{-1}$. At small shear rates ($\approx 1\,\mathrm{s}^{-1}$), RBCs tumble without significant deformation. Above a few $10\,\mathrm{s}^{-1}$, they 'tank-tread' and behave more like a liquid droplet [47]. Tank-treading denotes the rolling motion of the membrane about its interior without changing the external shape. At even higher shear rates, RBCs are strongly elongated and aligned with the flow, which reduces the viscosity further. Beyond $100\,\mathrm{s}^{-1}$, the viscosity does not change significantly [46].

In the past 50 years, it has been tried to measure the yield stress of blood as function of hematocrit. It has been suggested that it scales as $\sigma_\mathrm{y} \propto (\mathrm{Ht} - \mathrm{Ht_y})^3$ with $\mathrm{Ht_y}$ between 0.01 and 0.07 [61], i.e., a yield stress seems to be present at very small volume fractions. Since it also depends on the squared fibrinogen concentration [7], it is likely related to biochemical effects.

Although Schmid-Schönbein and Wells [47] assumed that the shear thinning behavior of blood may be related to the presence of a yield stress, it is not clear whether the yield stress—if it exists—has a physiological significance [7]. In general, experiments investigating the low-shear rate viscosity of blood are difficult to perform [46]. Charm and Kurland [62] reported residual artifacts caused by interactions of the RBCs with the rheometer surface. The yield stress values obtained experimentally by Picart et al. [63] strongly depend on the structure of the wall: For smooth walls, the stress at small shear rates was found to be an order of magnitude smaller than for rough walls. The results are not conclusive for basically two reasons: (i) Different definitions of yield stress in blood are used by different scientists. (ii) The yield stress seems to be a function of time and depends on the shearing history of the blood sample. The above observations indicate that the yield stress of blood at volume fractions $\leq 45\%$ is related to the aggregation of RBCs due to fibrinogen.

A large number of effects related to hemorheology are caused by confinement, i.e., the finite size of blood vessels. These effects have to be distinguished from the general shear thinning behavior of blood which can be observed in bulk systems as well. For blood vessels with diameters smaller than 30 µm, the particulate nature of the blood has to be taken into account explicitly [64]. Above, the flow resistance can be predicted using a two-phase continuum model with a central core and a cell-free layer of about 1.8 µm thickness [65]. This cell-free layer is caused by lift forces related to the presence of the walls and the deformability of the RBCs [66, 67]. Still, it is not clear how to predict the cell-free layer in general cases, e.g., for complex geometries and varying RBC properties [51]. One ultimate goal of hemorheology is to find reliable constitutive models for blood flow on the macroscale (both under physiological and pathological conditions), derived from its properties on the microscale. This becomes complicated due to the presence of different time and length scales (aggregate decay, cell deformation, etc.) [46]. For blood vessels with diameters larger than 200 µm, confinement becomes more and more negligible [68] and effective homogeneous models are commonly used. A large variety of constitutive models have been proposed by Truskey et al. [7] and Robertson et al. [46].

The cell-free layer plays an important role for biological transport, especially in blood vessels with diameters of about 10 µm [51]. The famous Fåhraeus effect [69] is tightly related to the existence of the cell-free layer: More than 80 years ago, Fåhraeus [70] investigated blood flow from a large feeding tube into a smaller tube with diameters between 0.05 and 1.5 mm. Due to the lateral migration of the RBCs towards the centerline, the cells move faster than the average flow. The mass balance then requires that the hematocrit in the small tube (tube hematocrit) is smaller than the hematocrit in the reservoir (discharge hematocrit). Another effect, closely connected to the cell-free layer, is the Fåhraeus-Lindqvist effect [71]. It states that the flow rate in small tubes for a given pressure gradient strongly depends on the tube diameter and that the apparent blood viscosity is minimum for a pipe diameter of about 10 µm. Velocity profiles of blood in tubes are usually characterized by flattening near the central region (plug flow) [72]. The reason is that the shear stress near the centerline vanishes whereas it is maximum close to the walls. Due to shear thinning, the viscosity of blood is highest near the centerline, and the curvature of the velocity profile is reduced. An additional effect is the inhomogeneous RBC concentration. Due to the existence of the cell-free layer, the viscosity of blood near the vessel walls is basically the viscosity of plasma, and the hematocrit is highest near the centerline of the flow.

Part II.

Numerical model for simulations of red blood cell suspensions

4. Physical considerations and ingredients for the numerical model

The physical requirements and ingredients for the computational model employed in the present thesis are characterized in this chapter. An overview of existing approaches for the simulation of individual and multiple deformable particles immersed in a fluid is provided in section 4.1. It is discussed in section 4.2 which physical ideas and concepts should be contained in the model and which can be disregarded.

4.1. Overview of existing numerical approaches

Suspensions of rigid spheres have been simulated by Ladd [73, 74] and Aidun and Lu [75] within the framework of the lattice Boltzmann method (LBM, chap. 5). In 1996, Kraus et al. [76] have simulated, for the first time, a single deformable vesicle in an external shear flow using the boundary integral method. Two years later, Eggleton and Popel [77] combined the immersed boundary method (IBM, cf. chap. 6) [78] and a finite element method in order to simulate deformable capsules. In 2004, Feng and Michaelides [79] were the first to combine the LBM with the IBM and simulated suspensions of rigid 2D disks. Zhang et al. [68, 80] also used a combination of the IBM and the LBM for red blood cell (RBC) simulations, but still in 2D. Even in 2005, Sun and Munn [81, 82] approximated RBCs and leukocytes as rigid particles in a 2D lattice Boltzmann simulation.

A number of articles about single RBC or vesicle dynamics in external flow fields has been published in the past ten years. Noguchi and Gompper [83] studied the effect of membrane viscosity on vesicle dynamics in shear flow, taking thermal fluctuations into account. The authors combined a dynamically triangulated membrane model with the multiparticle collision dynamics in 3D. Pozrikidis published a series of articles about the simulation of RBCs in shear flow via the boundary integral method (e.g., [84, 85]). Due to its computational overhead [84, 86], the boundary integral method seems to be not suitable for the simulation of a large number of RBCs.

The simulation of deformable RBC suspensions was promoted in 2007 when Dupin et al. [87] combined the LBM and the IBM with a spring model for the RBC membranes in 3D. 200 cells with a volume fraction of 30% could be simulated. However, a larger number of particles and a higher volume fraction was not obtainable at that time. In the same year, Bagchi [88] simulated 2500 RBCs in 2D. This model was extended to 3D by Doddi and Bagchi [89] two years later. MacMeccan et al. [90] simulated deformable RBCs via a lattice Boltzmann finite element method.

Concluding, a large variety of simulation methods for particle suspensions has been proposed in recent years. Some of the methods have been implemented for 2D only, others approximate the deformable particles as rigid objects. While some of the methods are of high accuracy and mostly suitable for a small number of particles (e.g., the boundary integral method), other approaches are less accurate but more efficient and simpler (e.g., the IBM). Still, neither of these methods seems to be able to combine all of the properties required for the study in the present work: (i) 3D simulations, (ii) deformable and resolved particles, (iii) volume fractions larger than 45%, and (iv) high runtime efficiency with $\mathcal{O}(10^3)$ particles.

4.2. Identification of the relevant physics for the present task

One of the main motivations of the present thesis is the development and application of a numerical tool for the simulation of dense suspensions of deformable particles, e.g., RBCs. Since the focus of the work lies on the investigation of collective phenomena, the ability to simulate a large number of particles is favored over high accuracy for only a few suspended objects. As a consequence, the single particle dynamics should be simplified as much as possible without losing the advantage of tracking the deformation of individual particles in the suspension. Although large progress has been made in the field of computational physics and computing power in recent years, large scale simulations of deformable particles have always required certain idealizations of the physics on the smallest resolved scales ($\approx 0.5\,\mu\mathrm{m}$ in the present case).

In dense suspensions, the immersed particles are no passive tracers comoving with the suspending fluid. Instead, the *bidirectional* influence of the fluid and the particles is one of the key factors for successful simulations of particle suspensions. In fact, Einstein's famous expression for the viscosity of dilute suspensions, 2.6, reflects that suspended particles, deformable or not, affect the fluid rheology, even in the dilute limit. The shear thinning behavior of blood at intermediate shear rates (a few $10\,\mathrm{s}^{-1}$), caused by the deformability of the RBCs [54], is another striking argument for the paramount importance of the bidirectional coupling of hydro- and particle dynamics. For that reason, the model has to be based on a two-way coupling: The fluid exerts stresses on the particle surfaces, and the presence of the surfaces poses a boundary condition for the fluid. Kraus et al. [76] formulated this in the following way:

> Any theory of vesicle dynamics is complicated by the fact that the boundary conditions for the three-dimensional Navier-Stokes equations have to be evaluated at the vesicle surface, which is moving with the fluid and whose shape is not known a priori.

This statement can directly be extended to any other kind of deformable particles immersed in a fluid. The IBM will be employed as efficient two-way fluid-structure coupling (chap. 6).

Since the rheology of deformable particles is of primary interest here, only the *mechanical* properties of the particles shall be considered. When RBCs are simulated, their biophysical and biochemical properties (e.g., aggregation at small shear rates or non-hydrodynamic interactions with other cells or the endothelium) are not taken into account. The deformable particles are considered as effective 2D membranes immersed in an ambient 3D fluid. A scale separation between the membrane thickness and the membrane diameter is assumed. For RBCs, this is an excellent approximation since the membrane thickness is 4 nm compared to $8\,\mu\mathrm{m}$ cell diameter [31]. Another simplification is to neglect the viscosity of the membranes. In the model, dissipation only takes place in the fluid (inside and outside of the particles), and the membranes are purely elastic. The main reason for this step is to reduce the complexity of the parameter space. Membrane viscosity may be added to the model in the future.

The deformability of RBCs is a key factor for the shear thinning behavior of blood at shear rates above a few s^{-1} [47, 54, 91]. For this reason, it shall be investigated how the deformability of suspended particles affects the viscosity of the suspension and the statistical motion of the particles. The model, therefore, should provide a controllable particle deformability. Physics happening on scales smaller than the spatial resolution of the simulations ($\approx 0.5\,\mu\mathrm{m}$) cannot be resolved explicitly and must be put in by hand as effective ingredients. This including the elastic model for the membranes which is a consequence of its nanometer scale structure. The model for the membrane physics will be presented in chap. 7.

Membrane rearrangements are ignored in the present work, i.e., neighboring points on the membrane will always remain neighbors. This way, the numerical model for the membranes is drastically simplified since the numerical mesh topology is preserved (section 8.3). Applied to the simulation of RBCs, this simplification is still reasonable because experimental investigations indicate that RBCs have a shape memory, i.e., the rim and the dimples of a RBC are always formed by the same patches of the membrane surface, even after deformations which are long compared to the typical advection times in the human body [92]. Thus, it can be assumed that membrane rearrangements are not important for RBCs, at least on time scales accessible by simulations.

The fluids both in the interior and the exterior of the particles are assumed to be Newtonian (which is also the case for RBCs [51, 93]). For the sake of computational efficiency, a single density and viscosity will be used for both the interior and the exterior fluids. The Newtonian fluid is modeled via the LBM as described in chap. 5. Due to the length scale separation between the fluid molecules and the RBCs, the suspending fluid does not need to be described on the kinetic level.

Thermal effects of any kind are neglected. On the one hand, a possible temperature dependent behavior of material parameters is ignored by assuming that the temperature is constant throughout the system and at all times (infinite heat conductivity). On the other hand, thermal fluctuations are not considered. The particle diameters considered in this thesis are a few micrometers or larger, thus, the particles can be considered non-Brownian [13, 94]. This is especially true for RBCs with diameters of about 8 μm. Additionally, for particles consisting of thin membranes (e.g., vesicles or RBCs), the thermal membrane fluctuations can be neglected when the energy scale for bending resistance is sufficiently large [31, 95].

The LBM (chap. 5) and the IBM (chap. 6) are efficient numerical tools. The membrane model as introduced in chap. 7 contains the physics relevant for the chosen length and energy scales, but it is not burdened with irrelevant and computationally expensive details. The resulting numerical model is highly efficient, which benefits the achievable system size and duration of the simulations.

5. Fluid solver: the lattice Boltzmann method

This chapter is dedicated to the lattice Boltzmann method (LBM) which is used as Navier-Stokes solver in the present work. In section 5.1, an introduction to the LBM is given. The employed LBGK algorithm, a special case of the LBM, is discussed in section 5.2. Initial and boundary conditions in general (section 5.3) and the bounce-back boundary condition in particular (section 5.4) are discussed. Section 5.5 deals with the choice of simulation parameters. Due to its technical character, the Chapman-Enskog analysis linking the mesoscopic LBGK and the macroscopic hydrodynamics is presented in appx. B.1, rather than in this chapter. A benchmark test for the LBM and its velocity and stress convergence is presented in appx. B.1.4.

5.1. Introduction to the lattice Boltzmann method

In 1986, the lattice gas cellular automata (LGCA) have been introduced by Frisch et al. [96] and Wolfram [97]. The basic idea was—in spirit of the Boltzmann equation—to model individual kinetic particles subject to propagation and collision on a hexagonal lattice in order to recover hydrodynamics on larger scales. Particles moving in a given direction can either be present or absent (Boolean arithmetics). However, there are some disadvantages which detract the applicability of the LGCA for hydrodynamic problems [98, 99]: The pressure of the fluid is velocity-dependent, and the momentum equation has an additional unphysical factor. Both artifacts are related to the violation of Galilei invariance by the LGCA algorithm. The viscosity of the LGCA fluid is intrinsically high, i.e., large Reynolds numbers and turbulence simulations are not feasible. Moreover, due to its statistical nature, the results show strong fluctuations. Spatial and temporal averaging is required to obtain smooth solutions. Another point is that the original lattice gas algorithm becomes extremely inefficient in 3D.

In order to remove the shortcomings of the LGCA, the LBM has been introduced in 1988. The intention was to keep the advantages of the LGCA (intrinsic stability, simple boundary handling, coding, and parallelization) at the same time. The difficulties in solving the Boltzmann equation originate from the complexity of the collision operator. Therefore, a lot of work has been spent on finding ways to reduce its complexity without losing the asymptotically hydrodynamic properties of the method. Important steps in the development of the LBM were (i) the transition from the Boolean to real number arithmetics eliminating the LGCA fluctuations [100], (ii) the linearization and enhancement of the LGCA collision operator making 3D simulations feasible [101, 102], and (iii) the combination of the Bhatnagar-Gross-Krook (BGK) collision operator [103] with the LBM [104, 105, 106]. The resulting algorithm is called lattice Bhatnagar-Gross-Krook (LBGK) method. It will be presented in section 5.2. In principle, the idea of the LBGK is to forget about all details of the kinetic theory which are not absolutely necessary to solve the Navier-Stokes equations (NSE).

Consequently, the LBGK is a numerical solver for the BGK approximation of the Boltzmann equation,

$$\frac{\partial f}{\partial t} + \boldsymbol{c} \cdot \nabla f = -\frac{1}{\tau'}(f - f^{\mathrm{eq}}), \tag{5.1}$$

where f is the local velocity distribution, $\boldsymbol{c}$ is the particle velocity, and τ' is a relaxation time.

The equilibrium Maxwell-Boltzmann distribution is given by

$$f^{\text{eq}} = \frac{\rho}{(2\pi RT)^{3/2}} \exp\left(-\frac{m(\boldsymbol{c}-\boldsymbol{u})^2}{2kT}\right) \tag{5.2}$$

with particle mass m and density $\rho := \int \mathrm{d}^3c\, f$, temperature T, the Boltzmann constant k, and the gas constant R. Here, one has to distinguish between the microscopic velocity $\boldsymbol{c}$ and the macroscopic velocity $\boldsymbol{u} := \int \mathrm{d}^3c\, f\boldsymbol{c}/\rho$. It has to be noted that the standard LBM describes an athermal system, i.e., a system with finite but constant temperature T in the absence of thermal fluctuations. The energy equation for heat flux is not considered. Instead, it is assumed that heat production by viscous dissipation is small and that the thermal conductivity of the fluid is sufficiently large to maintain a constant temperature.

It is important to emphasize that—although they asymptotically solve the NSE—neither the LBM in general, nor the LBGK in particular are direct discretizations of the NSE. Instead, they are based on mesoscopic models and kinetic equations. In fact, it is possible to recover the LBM through different approaches, e.g., directly from the Boltzmann equation or from discrete velocity models [107, 108, 109]. The Boltzmann equation describes gases both on the kinetic and the continuum scale. In order to recover the macroscopic Navier-Stokes behavior, small Knudsen numbers are necessary. This means that the fluid is assumed to be close to its local equilibrium everywhere, and that the non-equilibrium contributions can be treated perturbatively. This idea and the consequent derivation of the NSE from the LBGK formalism will be discussed in appx. B.1.

The LBM has a series of advantages compared to other Navier-Stokes solvers. First of all, due to its local kinetic scheme which the LBM has inherited from the LGCA, the LBM is intrinsically parallelizable. It reveals advantages especially when it comes to the simulation of multiphase and multicomponent flows [64, 79, 80]. When the incompressible NSE are directly solved, the pressure equation—which does not exist in the LBM—can lead to numerical difficulties [109]. Although finite element methods are very flexible in their application, they are difficult to code and require massive computational resources [110]. LBM coding is simple, at least for basic hydrodynamic problems. The LBM is also advantageous when particle suspensions are simulated. Due to the complete separation of time scales for the fluid and the particles in conventional Navier-Stokes solvers, the computational cost usually scales with the square or even the cube of the particle number N [73]. The LBM scales linearly with N since time scales are not exactly separated and fluid-particle interactions remain local. This is one of the major reasons to employ the LBM for the present work.

The LBM has been applied to various hydrodynamical problems such as flows in porous media [111, 112], dendritic growth [113], or turbulent flows [114, 115]. It has also been used to solve Maxwell's equations in materials [116] and for mildly relativistic hydrodynamics [117]. Further details about the LBM can be found in the literature: Beside monographs about the LBM by Wolf-Gladrow [118], Succi [98], and Sukop and Thorne [119], there exist various topical review articles, e.g., a general overview [109], a review of the LGCA and LBM with focus on the mathematical background [120], a review of the LBM for particle-fluid suspensions [110], a review focused on high Reynolds number flows in the LBM [121], an overview of LBM for materials science and engineering [122], and a review of the LBM for complex flows [12].

5.2. LBGK algorithm

The algorithm of the LBGK method is presented in this section. A Chapman-Enskog analysis linking the LBGK formalism with the macroscopic NSE is provided in appx. B.1.

While conventional methods directly solve the NSE in terms of the pressure p and the velocity $\boldsymbol{u}$, the LBM introduces a number of q populations $f_i(\boldsymbol{x},t)$ $(i=0,\ldots,q-1)$ streaming along a regular, d-dimensional lattice (lattice constant Δx) in discrete time steps (step size Δt). The population $f_i(\boldsymbol{x},t)$ can be regarded as the discretized probability distribution of finding a particle at position $\boldsymbol{x}$ and time t, moving with velocity $\boldsymbol{c}_i$. In the LBM, the velocity space is also discretized, allowing only for q possible velocity states $\boldsymbol{c}_i$.

The evolution of the populations in the BGK approximation is given by the LBGK equation,

$$f_i(\boldsymbol{x}+\boldsymbol{c}_i\Delta t, t+\Delta t) - f_i(\boldsymbol{x},t) = \Omega_i(\boldsymbol{x},t), \tag{5.3}$$

where

$$\Omega_i(\boldsymbol{x},t) = -\frac{1}{\tau}\left(f_i(\boldsymbol{x},t) - f_i^{\mathrm{eq}}(\boldsymbol{x},t)\right) + f_i^F(\boldsymbol{x},t)\Delta t \tag{5.4}$$

is the linearized collision operator. The idea is to let all populations relax towards a local equilibrium, f_i^{eq}, with a constant and unique relaxation time $\tau\Delta t$ where τ is the dimensionless relaxation parameter. The f_i^F are used to add external forces (see below). This algorithm is also called the single relaxation time approximation of the more general quasi-linear LB equation which reads

$$f_i(\boldsymbol{x}+\boldsymbol{c}_i\Delta t, t+\Delta t) - f_i(\boldsymbol{x},t) = A_{ij}\left(f_j(\boldsymbol{x},t) - f_j^{\mathrm{eq}}(\boldsymbol{x},t)\right) \tag{5.5}$$

in the absence of external forcing. (A_{ij}) is called the scattering matrix, and obviously $A_{ij}=-\delta_{ij}/\tau$ holds for the LBGK approximation. 'Quasi-linear' means that the collision term is formally linear in the populations f_i whereas the non-linearity is hidden in the equilibrium populations f_i^{eq} which will be defined soon. Interestingly, this implicit non-linearity allows to recover the non-linear NSE from the strongly simplified LBGK scheme (appx. B.1).

At each time step t, the populations collide and relax at the fixed lattice nodes $\boldsymbol{x}$ according to the right-hand-side of eq. (5.3). Afterwards, the populations propagate along the q discretized velocity vectors $\boldsymbol{c}_i$ to the next neighbors according to the left-hand-side of eq. (5.3). The equilibrium populations are given by

$$f_i^{\mathrm{eq}} = w_i\,\rho\left(1+\frac{\boldsymbol{c}_i\cdot\boldsymbol{u}}{c_{\mathrm{s}}^2}+\frac{\boldsymbol{Q}_i:\boldsymbol{u}\boldsymbol{u}}{2c_{\mathrm{s}}^4}\right) \tag{5.6}$$

where ρ and $\boldsymbol{u}$ are the fluid density and velocity, respectively, and $\boldsymbol{Q}_i := \boldsymbol{c}_i\boldsymbol{c}_i - c_{\mathrm{s}}^2\boldsymbol{I}$ is the velocity projector with unit matrix $\boldsymbol{I}$. Eq. (5.6) forms the truncated expansion of the Maxwell-Boltzmann distribution for the velocities in an ideal gas, eq. (5.2), with $c_{\mathrm{s}}^2 = kT/m$. The quantity c_{s} in eq. (5.6) is called the lattice speed of sound. Due to the truncation, eq. (5.6) is only valid for small Mach numbers[1], i.e., $|\boldsymbol{u}|$ must be sufficiently small compared to c_{s}.

The dimensionless relaxation parameter τ is connected to the lattice speed of sound c_{s} and the kinematic viscosity ν of the fluid by

$$\nu = c_{\mathrm{s}}^2\left(\tau-\frac{1}{2}\right)\Delta t. \tag{5.7}$$

Compared to the viscosity of the BGK gas, the LBGK viscosity has an additional term $-1/2$ which is a discretization artifact as will be shown in appx. B.1.1.

The lattices employed for the LBM are usually multispeed lattices, i.e., not all lattice velocities $\boldsymbol{c}_i$ have the same magnitude, cf. fig. 5.1. In order to account for this, it is necessary to introduce q weighting factors w_i which are uniquely defined for each underlying lattice. Their choice ensures

[1]The lattice Mach number is defined as $\mathrm{Ma} = |\boldsymbol{u}|/c_{\mathrm{s}}$.

the isotropy of the lattice, a necessity to solve the NSE asymptotically. It can be shown that the lattice weights have to obey the following equations [123]:

$$\begin{aligned}
&\sum_i w_i = 1, && \sum_i w_i c_{i\alpha} = 0, \\
&\sum_i w_i c_{i\alpha} c_{i\beta} = c_\mathrm{s}^2 \delta_{\alpha\beta}, && \sum_i w_i c_{i\alpha} c_{i\beta} c_{i\gamma} = 0, \\
&\sum_i w_i c_{i\alpha} c_{i\beta} c_{i\gamma} c_{i\kappa} = c_\mathrm{s}^4 (\delta_{\alpha\beta}\delta_{\gamma\kappa} + \delta_{\alpha\gamma}\delta_{\beta\kappa} + \delta_{\alpha\kappa}\delta_{\beta\gamma}), && \sum_i w_i c_{i\alpha} c_{i\beta} c_{i\gamma} c_{i\kappa} c_{i\lambda} = 0.
\end{aligned} \tag{5.8}$$

Here and in the following, the sum over lattice velocity indices i always runs from 0 to $q-1$ if not otherwise stated.

For low Reynolds number flows, it is possible to use the linearized equilibrium populations [73]

$$f_i^\mathrm{eq} = w_i \, \rho \left(1 + \frac{\boldsymbol{c}_i \cdot \boldsymbol{u}}{c_\mathrm{s}^2}\right) \tag{5.9}$$

instead of eq. (5.6). As will be shown in appx. B.1.1, this leads to the NSE without the advective term $\rho(\boldsymbol{u} \cdot \nabla)\boldsymbol{u}$.

It is shown in appx. B.1.1 that the fluid described by the LBGK equation is not exactly incompressible. Instead, the equation of state is

$$p = c_\mathrm{s}^2 \rho, \tag{5.10}$$

i.e., that of an ideal gas ($p \propto \rho$ for constant temperature), where p and ρ are the pressure and the density of the fluid, respectively. The slight compressibility of the LBM is the prize for its locality since the kinetic and hydrodynamic time scales are not exactly separated [73]. However, in practical simulations and for sufficiently small pressure gradients and Mach numbers, the fluid can be considered incompressible.

As already mentioned above, a body force density $\boldsymbol{f}$ (force per volume) may be incorporated via f_i^F in eq. (5.3) [110, 124],

$$f_i^F = \left(1 - \frac{1}{2\tau}\right) w_i \left(\frac{\boldsymbol{c}_i - \boldsymbol{u}}{c_\mathrm{s}^2} + \frac{\boldsymbol{c}_i \cdot \boldsymbol{u}}{c_\mathrm{s}^4} \boldsymbol{c}_i\right) \cdot \boldsymbol{f}. \tag{5.11}$$

This force density is particularly important for the introduction of gravity or the coupling of the fluid and immersed particles (chap. 6).

For most of the simulations in this thesis, if not otherwise stated, the 3D model with 19 velocities, D3Q19 ($d = 3$, $q = 19$), is used. A sketch of the D3Q19 lattice is shown in fig. 5.1. In the benchmark presented in appx. B.1.4, the D2Q9 lattice is employed. For both lattices[2], the speed of sound is $c_\mathrm{s} = \sqrt{1/3}\,\Delta x/\Delta t$. It should be noted that the discussions about the LBGK algorithm in this section and in appx. B.1 are equally valid for D3Q19 and D2Q9.

For D3Q19, the lattice weights are defined as $w_0 = 1/3$ (zero velocity), $w_{1\ldots 6} = 1/18$ (next neighbors), and $w_{7\ldots 18} = 1/36$ (next but one neighbors). The 19 lattice velocities read

$$(\boldsymbol{c}_i) = \left(\begin{array}{c|cccccc|cccccccccccc}
0 & 1 & -1 & 0 & 0 & 0 & 0 & 1 & -1 & -1 & 1 & -1 & 1 & 1 & -1 & 0 & 0 & 0 & 0 \\
0 & 0 & 0 & 1 & -1 & 0 & 0 & 0 & 0 & -1 & 1 & 0 & 0 & -1 & 1 & -1 & 1 & -1 & 1 \\
0 & 0 & 0 & 0 & 0 & 1 & -1 & 1 & -1 & 0 & 0 & 1 & -1 & 0 & 0 & 1 & -1 & -1 & 1
\end{array}\right) \frac{\Delta x}{\Delta t}. \tag{5.12}$$

[2]The value of the lattice speed of sound depends on the chosen lattice. At least for D1Q3, D2Q9, D3Q15, D3Q19, and D4Q25, it is $\sqrt{1/3}\,\Delta x/\Delta t$ [106].

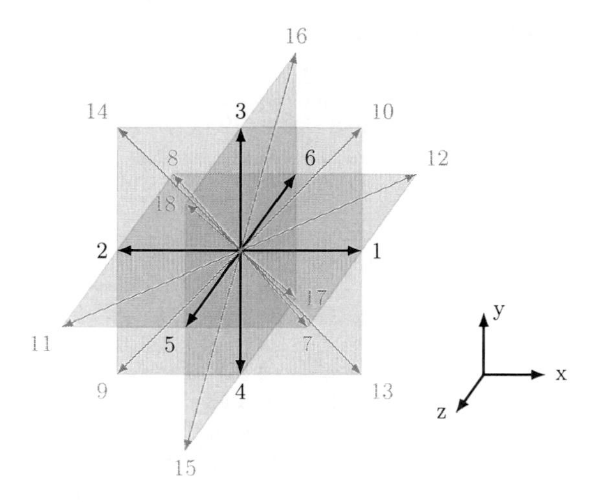

Fig. 5.1.: D3Q19 lattice for the lattice Boltzmann method. For D3Q19, all velocity vectors $\boldsymbol{c}_i$ are located in at least one of the three coordinate planes (light blue). The velocity vectors either point to the next neighbors along the coordinate axes ($\boldsymbol{c}_{1-6}$, black arrows) or to the next but one neighbors ($\boldsymbol{c}_{7-18}$, gray arrows). There are no body diagonals with length $\sqrt{3}\Delta x$, cf. eq. (5.12). The zero velocity $\boldsymbol{c}_0$ is not shown.

For the D2Q9 lattice, the weights are $w_0 = 4/9$, $w_{1...4} = 1/9$, $w_{5...8} = 1/36$, and the velocities read

$$(\boldsymbol{c}_i) = \left(\begin{array}{c|cccc|cccc} 0 & 1 & -1 & 0 & 0 & 1 & -1 & 1 & -1 \\ 0 & 0 & 0 & 1 & -1 & 1 & -1 & -1 & 1 \end{array} \right) \frac{\Delta x}{\Delta t}. \tag{5.13}$$

An overview of these and other lattices is also given in [106].

The fluid density and velocity can be obtained from the first two moments of the populations,

$$\rho = \sum_i f_i, \tag{5.14}$$

$$\rho \boldsymbol{u} = \sum_i f_i \boldsymbol{c}_i + \frac{\Delta t}{2} \boldsymbol{f} \tag{5.15}$$

where the second term on the right-hand-side of eq. (5.15) is a lattice correction term [110, 124], cf. appx. B.1.1. The deviatoric stress tensor $\boldsymbol{\sigma}$ can be extracted locally from the non equilibrium populations

$$f_i^{\text{neq}} := f_i - f_i^{\text{eq}} \tag{5.16}$$

by computing the second moment

$$\boldsymbol{\sigma} = -\left(1 - \frac{1}{2\tau}\right) \sum_i f_i^{\text{neq}} \boldsymbol{c}_i \boldsymbol{c}_i + \boldsymbol{X}_{\text{c}}. \tag{5.17}$$

Here, $\boldsymbol{X}_{\text{c}}$ is a lattice correction term [124, 125]. It is shown in appx. B.1.1 that, for the quadratic equilibrium in eq. (5.6), it reads

$$\boldsymbol{X}_{\text{c}} = -\frac{\Delta t}{2} \left(1 - \frac{1}{2\tau}\right) (\boldsymbol{u}\boldsymbol{f} + \boldsymbol{f}\boldsymbol{u}). \tag{5.18}$$

Contrarily, for the linearized equilibrium in eq. (5.9), $\boldsymbol{X}_{\text{c}}$ vanishes. The deviatoric stress tensor is of paramount importance for the study of suspension rheology as will be argued in chap. 9.

5.3. Initial and boundary conditions in the lattice Boltzmann method

The NSE, being a system of partial differential equations, cannot be solved without specifying initial and boundary conditions. The same applies for the LBGK equation. Since the LBGK equation is used to solve the NSE asymptotically, the corresponding macroscopic initial and boundary conditions (on the Navier-Stokes level) have to be translated to the mesoscale (on the lattice Boltzmann level). In the following, brief discussions of the initial and boundary conditions in the LBM are given.

5.3.1. Initial conditions in the lattice Boltzmann method

In many practical applications of hydrodynamics at small Reynolds numbers, the initial conditions do not play an important role. For steady flows, the flow field usually converges to the steady solution after some transient time, even if the initial conditions are not appropriately chosen. The reason is the viscous dissipation which erases the memory of the fluid. For unsteady flows at higher Reynolds numbers and, especially, for the study of transient effects, the initial conditions are relevant.

If the initial pressure p, velocity $\boldsymbol{u}$, and viscous stress $\boldsymbol{\sigma}$ are known, the populations f_i can be initialized according to

$$f_i = f_i^{\mathrm{eq}} + f_i^{\mathrm{neq}} \tag{5.19}$$

where the equilibrium populations $f_i^{\mathrm{eq}}(\rho, \boldsymbol{u})$ are given in eq. (5.6) or eq. (5.9) and the non-equilibrium populations, f_i^{neq}, eq. (5.16), may be reconstructed from the stress tensor $\boldsymbol{\sigma}$ via [126]

$$f_i^{\mathrm{neq}}(\boldsymbol{\sigma}) = -\frac{w_i}{2c_{\mathrm{s}}^4}\frac{1}{1-\frac{1}{2\tau}}\boldsymbol{Q}_i : \boldsymbol{\sigma} \tag{5.20}$$

if there is *no* forcing involved. It has to be noted that the pressure and the density are coupled via $p = c_{\mathrm{s}}^2\rho$, cf. section 5.2 and appx. B.1. Similar to the computation of the equilibrium populations, the idea behind eq. (5.20) is to find a local expression for the non-equilibrium populations which is a pure function of the macroscopic observables and which is self-consistent, i.e., the correct density, velocity, and stress must be recovered from these populations. It should be mentioned that eq. (5.20) is not the only possible expression for a self-consistent set of non-equilibrium populations [126].

Corrections due to forcing as indicated in eq. (5.15) and eq. (5.17) are not considered in eq. (5.20). It is shown in appx. B.2 that eq. (5.20) has to be modified in order to account for forcing: for the quadratic equilibrium in eq. (5.6),

$$f_i^{\mathrm{neq}}(\boldsymbol{u}, \boldsymbol{\sigma}, \boldsymbol{f}) = -\frac{w_i}{2c_{\mathrm{s}}^4}\frac{1}{1-\frac{1}{2\tau}}\boldsymbol{Q}_i : \boldsymbol{\sigma} - \frac{w_i\Delta t}{2c_{\mathrm{s}}^2}\boldsymbol{c}_i \cdot \boldsymbol{f} - \frac{w_i\Delta t}{4c_{\mathrm{s}}^4}\boldsymbol{Q}_i : (\boldsymbol{u}\boldsymbol{f} + \boldsymbol{f}\boldsymbol{u}) \tag{5.21}$$

and for the linearized equilibrium in eq. (5.9),

$$f_i^{\mathrm{neq}}(\boldsymbol{u}, \boldsymbol{\sigma}, \boldsymbol{f}) = -\frac{w_i}{2c_{\mathrm{s}}^4}\frac{1}{1-\frac{1}{2\tau}}\boldsymbol{Q}_i : \boldsymbol{\sigma} - \frac{w_i\Delta t}{2c_{\mathrm{s}}^2}\boldsymbol{c}_i \cdot \boldsymbol{f}. \tag{5.22}$$

Neglecting the non-equilibrium populations in eq. (5.19) may lead to an inaccurate initial configuration detrimentally affecting the entire time evolution of the simulation. For unsteady flows evolving from a non-trivial[3] initial configuration, the non-equilibrium populations should

[3]Trivial means initially constant pressure, zero velocity, zero stress, and zero deviatoric stress.

always be considered as well. If the initial conditions are not known in a closed form, it is usually necessary to solve a Poisson equation first.

Additional details about initial conditions in the LBM can only be found in a few references, e.g., [126, 127, 128]. In the present thesis, trivial initial conditions (constant density, zero velocity and force) are used if not otherwise stated, and eq. (5.19) is employed for initialization.

5.3.2. Boundary conditions in the lattice Boltzmann method

In hydrodynamics, boundary conditions (BCs) are usually posed in terms of the velocity or pressure (Dirichlet BCs), velocity or pressure gradient (Neumann BCs), or a combination of both (Robin BCs). The LBM does not directly solve the NSE in terms of the velocity $\boldsymbol{u}$ and pressure p. Instead, the evolution of the populations f_i is described, and the velocity and pressure fields are recovered from the populations (section 5.2). As a consequence, in the LBM, the BCs have to be formulated for the populations, either from hydrodynamic considerations or from a kinetic approach. However, while it is straightforward to compute the hydrodynamic variables from the populations, it is not directly obvious how to reconstruct the populations from the macroscopic variables. This is, actually, the reason for the large number of works which have been published about the implementation of LBM BCs in the recent years.

In computational fluid dynamics, periodic BCs significantly simplify the boundary treatment since the fluid domain is not bounded by walls and no populations have to be reconstructed in the LBM. It is obvious that general flow configurations cannot be mimicked by periodic BCs, and the inclusion of plain walls or more complex bounding geometries remains inevitable in many cases.

One of the most common BCs in hydrodynamic applications is the no-slip BC. It is a special case of a Dirichlet BC for the velocity. The basic idea is that the fluid velocity at a solid wall equals the velocity of the wall at the contact point, i.e., that there is no wall slip. The microscopic origin of the no-slip behavior has not been established with certainty [129]. However, it is commonly argued that fluid molecules are adsorbed by the wall for a short time and desorbed afterwards. This interaction slows down the particle dynamics close to the wall leading to an effective no-slip behavior. The assumption of no-slip in hydrodynamic flows seems to be valid for flow geometries larger than a few 10 nm [130].

A particularly simple and efficient way to enforce no-slip BCs at solid walls is the bounce-back (BB) BC. Due to its importance for the present thesis, it is detailedly explained in section 5.4.

Throughout this thesis, mostly periodic and BB BCs are used. However, since the problem of BCs in the LBM has an enormous significance, a short overview of alternative methods is provided in the following.

The basic idea of BCs in the LBM is to obtain expressions for the equilibrium and non-equilibrium populations near or at the boundaries. While populations leaving the computational domain are not a problem, the unknown populations entering the domain from the outside have to be chosen in an appropriate way, cf. fig. 5.2. The implementation of a given BC should fulfill a series of requirements:

- The desired values for the hydrodynamic observables must be recovered, the velocity and pressure as well as the viscous stress.
- The BC must not violate the overall second order convergence of the LBM.
- The BC should be local.
- Local mass conservation should be maintained.

While the first two claims have to be enforced rigorously, the last two are less stringent and often violated in practical applications.

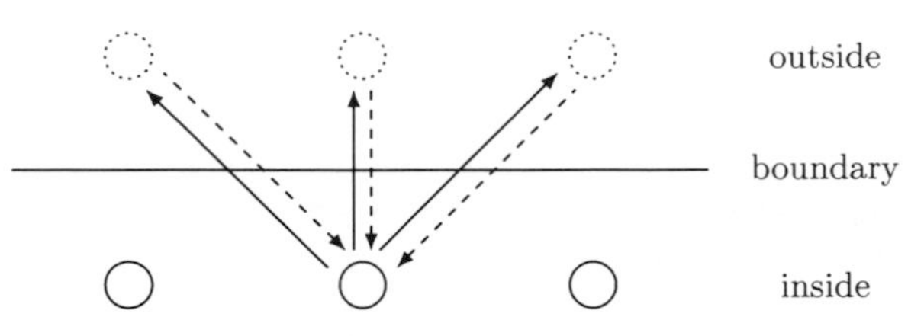

Fig. 5.2.: Unknown populations at boundaries in the lattice Boltzmann method. The black line separates fluid nodes (circles) inside and ghost nodes (dotted circles) outside of the computational domain. Populations leaving the domain (solid arrows) are known, populations entering the domain (dashed arrows) are not known a priori. The boundary conditions are required to find values for these populations in a self-consistent way.

A significant error source in LBM simulations stems from the BCs [131]. Thus—similar to the initialization in section 5.3.1—the correct reconstruction of the non-equilibrium populations by the BC schemes is of utmost importance. If the non-equilibrium populations are neglected, the velocity gradient is imposed in a wrong way, and the resulting errors may propagate throughout the entire volume and spoil the outcome of the simulation [132]. It turns out that the system of equations for the values of the equilibrium and non-equilibrium populations at boundary nodes is usually under-determined, which gives rise to the introduction of closure relations. These relations have a certain freedom, which is the reason for the vast variety of different BC schemes for the LBM in the literature. One possible way to reconstruct the non-equilibrium populations is to employ eq. (5.21) or eq. (5.22) where the stress tensor may be found by extrapolation from the bulk region [127]. Another approach is to guess a stress tensor from the known, outgoing non-equilibrium populations as thoroughly discussed by Latt et al. [132]. It is also possible to employ the 'bounce-back of non-equilibrium populations' to reconstruct the missing populations [133].

While the BB BC (section 5.4) is used for walls located half-way between lattice nodes, the majority of BCs in the LBM is tailored for walls being located directly on the nodes. Since these wall nodes—similar to the fluid nodes in the bulk—also participate in collision and propagation, those BCs are called 'wet' BCs (fig. 5.3). They often suffer from a violation of the local mass conservation [134].

The velocity and pressure BCs by Inamuro et al. [135] (based on the kinetic theory of gases) and Zou and He [133] (based on the lattice symmetry) are cornerstones in the research field. Latt [126, 132] introduced the regularized BC for straight walls which shares similarities with the approach by Skordos [127] involving non-local finite difference approximations for the recovery of the unknown non-equilibrium populations. Latt's BCs have been extended to curved walls by Verschaeve and Müller [136]. Other, more complex BCs are available, e.g., average outlet pressure in curved geometries [137], a velocity BC for arbitrarily shaped inlets [138], open boundaries [139, 140], slip-flow boundaries [141], macroscopic-gradient boundary conditions [142], periodic pressure boundaries [143], and Lees-Edwards BCs [144, 145]. Various existing BCs have been extensively reviewed by Latt et al. [132]. A recent overview of BCs is also provided by Izquierdo and Fueyo [142].

It is important to state that BCs can also be imposed in a different way: Instead of reconstructing the populations such that the BC is satisfied, a forcing term $\boldsymbol{f}(\boldsymbol{x}, t)$ may be added to mimic the local effect of the wall on the fluid. The immersed boundary method (IBM) sets exactly at this point (chap. 6). Due to its importance for the present thesis, it is thoroughly discussed in a separate chapter. Since the IBM affects the fluid only through a forcing term (and not by direct manipulation of the populations), it can also be employed for conventional Navier-Stokes solvers.

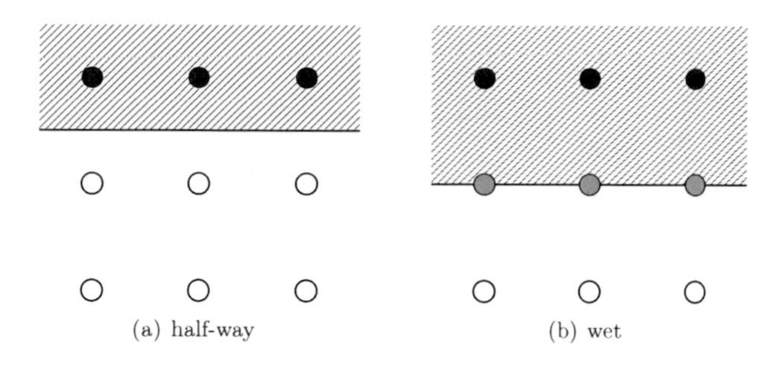

(a) half-way (b) wet

Fig. 5.3.: Comparison of half-way and wet boundary conditions in the lattice Boltzmann method. The physical wall (black line) separates the fluid region (white) from the obstacle region (striped). (a) Half-way boundary condition: The physical wall is located half-way between the fluid nodes (white circles) and obstacle nodes (black circles) as in the case of the bounce-back boundary condition. (b) Wet boundary condition: The physical wall is located directly on the wall nodes (gray circles). While the obstacle nodes (black circles) do neither participate in collision nor propagation, wall nodes have the same dynamics as fluid nodes, which gives rise to the name 'wet' boundary condition.

5.4. Bounce-back boundary condition

The idea of the BB BC goes back to the LGCA [146]. Populations hitting a solid wall during propagation are bounced back in the direction they came from [147]. This microscopic boundary rule leads to an asymptotic no-slip behavior of the macroscopic fluid at the wall [146]. The exact location of the physical wall, however, was a matter of debate for some time. Initially, the wall location was taken at the position of the node where populations are bounced back. It has been realized that this interpretation leads only to first-order accuracy, thus violating the overall second-order convergence. Later, it was found that the BB BC is second-order accurate for plane walls aligned with the lattice [110, 148, 149] if the physical wall is located half-way between the fluid and the wall node [150] (fig. 5.5).

The original BB BC is straightforward to implement, and the numerical overhead is negligible. Due to its simplicity, highly complex geometries (e.g., porous media) can be realized efficiently with the BB approach, an advantage over most of the other Navier-Stokes solvers where delicate meshing is required [73]. The disadvantage of the original BB BC is that curved boundaries are approximated by staircases leading to a reduction of the accuracy (fig. 5.4). It can even be reduced to first order in these cases [131, 147]. In applications where computational efficiency and simplicity of coding shall outweigh the accuracy, the BB is an outstanding candidate for solid wall BCs.

There are extensions of the BB scheme for inclined and curved walls (e.g., [148, 151, 152] and a review in [12]). However, these improvements are not discussed here since they are not used in the present thesis. Instead, the BB approach is mostly employed to mimic straight and rigid walls aligned with the lattice. In sections 5.4.1 and 5.4.2, it is discussed how the BB can be used to realize velocity and—apparently for the first time in the present form—shear stress BCs in the LBM.

5.4.1. Velocity bounce-back boundary condition

In section 5.3, it has already been discussed that BCs for the LBM eventually have to be formulated for the populations f_i in a self-consistent way. In particular, populations entering the fluid from the inside of a wall or from the outside of the numerical grid should be constructed in such a way that the macroscopic BC is satisfied and that the basic principles of the LBM

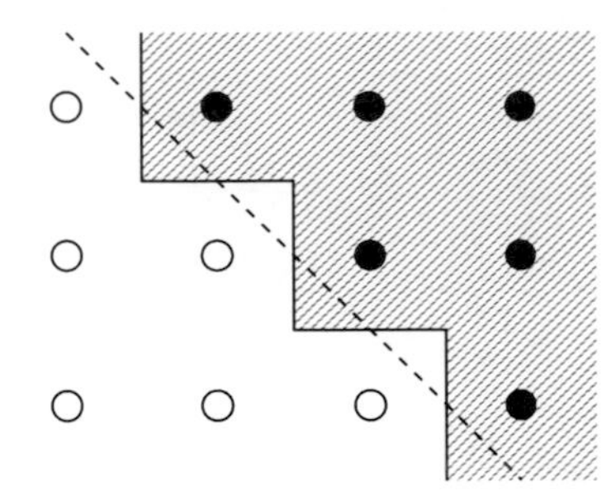

Fig. 5.4.: Staircase shape of an inclined wall in the lattice Boltzmann method. The staircase wall (solid black line) as approximation of the physical wall (dashed black line) separates the fluid region (white) with the fluid nodes (white circles) from the obstacle region (striped) with the obstacle nodes (black circles).

(locality, mass conservation, and second-order accuracy in space) are not violated. As will be shown in the following, these considerations are straightforward when BB BCs are used.

In order to better understand the idea behind BB, it is useful to distinguish between the pre-collision, post-collision, and post-propagation populations. The populations collide at the fluid nodes and then propagate to their next neighbors along the lattice velocity vectors $\boldsymbol{c}_i$. The BB scheme defines rules for the case that a population hits a wall while it propagates. In lattice units, the LBE from eq. (5.3) can be written in the form

$$f_i^*(\boldsymbol{x},t) = f_i(\boldsymbol{x},t) + \Omega_i(\boldsymbol{x},t), \qquad \text{(collision)} \qquad (5.23)$$

$$f_i(\boldsymbol{x}+\boldsymbol{c}_i,t+1) = f_i^*(\boldsymbol{x},t) \qquad \text{(propagation)} \qquad (5.24)$$

where Ω_i is the collision operator. A possible forcing term is absorbed in Ω_i and has no further significance for the following discussion. The pre-collision and post-propagation populations are denoted $f_i(\boldsymbol{x},t)$ and $f_i(\boldsymbol{x}+\boldsymbol{c}_i,t+1)$, respectively. The populations $f_i^*(\boldsymbol{x},t)$ are called the post-collision populations which have not been propagated yet. In case of BB, the collision at fluid nodes in eq. (5.23) is not changed, but the propagation in eq. (5.24) has to be modified: If the propagating populations $f_i^*(\boldsymbol{x},t)$ hit a resting wall (velocity $\boldsymbol{u}_\text{w} = 0$) at position $\boldsymbol{x} + \frac{1}{2}\boldsymbol{c}_i$ and time $t + \frac{1}{2}$, they are bounced back in the direction they came from, i.e., they propagate to the same fluid node where they were released before, and eq. (5.24) becomes

$$f_{i'}(\boldsymbol{x},t+1) = f_i^*(\boldsymbol{x},t). \qquad \text{(bounce-back)} \qquad (5.25)$$

During this process (fig. 5.5), the index of the bounced back populations changes from i to i' where i' denotes the index opposite to i, i.e., $\boldsymbol{c}_{i'} = -\boldsymbol{c}_i$. This automatically solves the problem of how to find the a priori unknown population $f_{i'}^*(\boldsymbol{x}+\boldsymbol{c}_i,t)$ apparently coming out of the wall. It can be shown that this algorithm recovers the behavior of the macroscopic fluid at a wall with zero velocity at position $\boldsymbol{x} + \frac{1}{2}\boldsymbol{c}_i$, i.e., the wall is located half-way between the lattice nodes. In this case, the BB BC is second-order accurate in space and thus fits into the overall lattice Boltzmann scheme [153]. Obviously, eq. (5.25) ensures mass conservation, and it is a local rule which makes the BB efficient in terms of implementation effort and computing time.

An important extension of the BB algorithm is the treatment of walls moving with non-zero velocity $\boldsymbol{u}_\text{w}$. Ladd [73] and Ladd and Verberg [110] thoroughly discussed this concern, and the modification of eq. (5.25) reads

$$f_{i'}(\boldsymbol{x},t+1) = f_i^*(\boldsymbol{x},t) - 2w_i\rho(\boldsymbol{x},t)\frac{\boldsymbol{u}_\text{w}\cdot\boldsymbol{c}_i}{c_\text{s}^2}. \qquad (5.26)$$

The new term on the right-hand side of eq. (5.26) captures the additional momentum transfer due to the motion of the wall. It can be derived via a Galilean transform to the coordinate

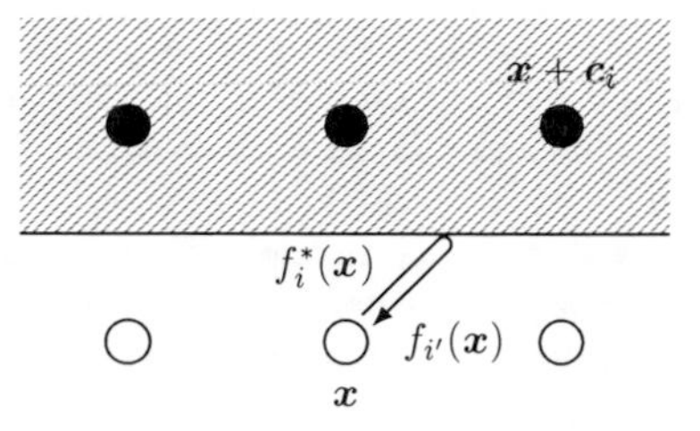

Fig. 5.5.: Bounce-back boundary condition in the lattice Boltzmann method. The physical wall (black line) separates the fluid region (white) with the fluid nodes (white circles) from the obstacle region (striped) with the obstacle nodes (black circles). For simplicity, the bounce-back of only one population is shown. The post-collision population $f_i^*(\boldsymbol{x}, t)$ propagates towards the obstacle node at $\boldsymbol{x} + \boldsymbol{c}_i$ but hits the wall at $\boldsymbol{x} + \frac{1}{2}\boldsymbol{c}_i$ at time $t + \frac{1}{2}$ before. It is bounced back and reaches its starting point as post-propagation population $f_{i'}(\boldsymbol{x}, t+1)$.

system where the wall is resting [144, 148]. Obviously, eq. (5.25) is recovered for $\boldsymbol{u}_\mathrm{w} = 0$. It should be noted that eq. (5.26) may lead to violation of mass conservation for general geometries if $\boldsymbol{u}_\mathrm{w} \neq \boldsymbol{0}$ [75]. However, for a straight wall moving in its own plane as in the present thesis, it can be inferred that the mass is always exactly conserved.

It has been stated that the BB BC is exact for a simple shear flow with a linear velocity profile [150]. However, for non-linear velocity profiles, the exact wall location is shifted with respect to the midplane between the lattice nodes, which introduces a finite slip velocity. Still being second-order accurate, the associated error becomes severe when the relaxation time is significantly larger than unity, and it grows without limits for $\tau \to \infty$.

Eq. (5.25) and eq. (5.26) can also be used for walls which are not straight but still obey the condition that they are located half-way between lattice nodes. As mentioned before, this leads to a staircase shape of the obstacle (fig. 5.4). It is the simplest way to approximate inclined walls or other complex geometries (e.g., spheres or porous media) and is often used in the literature (e.g., [145, 154, 155, 156, 157]). The numerical accuracy may be reduced in these cases, though [158].

5.4.2. Shear stress bounce-back boundary condition

The most common BCs in lattice Boltzmann simulations are periodic, pressure, or velocity BCs (section 5.3). In most applications, especially for simple fluids of well-known viscosity, this is satisfactory. Yet, there are also cases in which another type of BC is of advantage. In order to measure the static yield stress of complex fluids, it is necessary to apply a given shear stress to the system, not a shear rate. The reason for this is that soft materials in the glassy state do not generally show the same rheological response when driven by an imposed shear rate or shear stress [21]. For example, in general, the static yield stress is higher than the dynamic yield stress [20] where the former can be inferred from shear stress and the latter from shear rate driven experiments or simulations. To this end, it is desirable to implement, within the lattice Boltzmann formalism, a shear stress BC. Actually, there are ways to enforce macroscopic gradient BCs [142]. This can be used to implement a shear stress BC *if* the viscosity of the fluid is known (since $\boldsymbol{\sigma} = 2\eta\boldsymbol{S}$ where $\boldsymbol{\sigma}$ and $\boldsymbol{S}$ are the stress and strain rate tensors, respectively). However, if the viscosity of the fluid is not known a priori (as it is the case for complex fluids), a velocity gradient BC may not recover the desired wall stress. However, it turns out that there exists a straightforward approach for a shear stress BC for straight walls which is based on the velocity BB BC in eq. (5.26) and eq. (5.27). This method apparently has not been considered in the literature so far. It will be described in the following for a system consisting of a fluid between two straight and parallel walls.

The wall shear stress (WSS) is of major importance in many hydrodynamic situations, especially for the circulatory system where it plays a dominating role for the vascular response to hemodynamics [46, 51, 53] or blood clotting and platelet adhesion [159, 160, 161, 162]. The WSS is the measure of the momentum transfered from the fluid to the wall[4]. Within the BB approach, it is straightforward to evaluate the wall stress via the momentum exchange of the populations hitting the wall. Each population f_i propagating from position $\boldsymbol{x}$ to the wall location $\boldsymbol{x}_\mathrm{w} = \boldsymbol{x} + \frac{1}{2}\boldsymbol{c}_i$, where it is bounced back, contributes to the momentum transfer with [73]

$$\begin{aligned}\Delta \boldsymbol{p}_i(\boldsymbol{x}_\mathrm{w}, t+\tfrac{1}{2}) &= \left(f_{i'}(\boldsymbol{x}, t+1) + f_i^*(\boldsymbol{x}, t)\right) \boldsymbol{c}_i \\ &= 2\left(f_i^*(\boldsymbol{x}, t) - w_i \rho(\boldsymbol{x}, t) \frac{\boldsymbol{u}_\mathrm{w}(\boldsymbol{x}_\mathrm{w}, t+\frac{1}{2}) \cdot \boldsymbol{c}_i}{c_\mathrm{s}^2}\right) \boldsymbol{c}_i\end{aligned} \tag{5.27}$$

where $\boldsymbol{u}_\mathrm{w}(\boldsymbol{x}_\mathrm{w}, t+\frac{1}{2})$ is the velocity of the wall at position $\boldsymbol{x}_\mathrm{w}$ and time $t+\frac{1}{2}$. Since the population can be considered to be located at $\boldsymbol{x}_\mathrm{w}$ at time $t+\frac{1}{2}$, the momentum exchange $\Delta \boldsymbol{p}_i$ is formally taken at $\boldsymbol{x}_\mathrm{w}$ and $t+\frac{1}{2}$. The last equality in eq. (5.27) follows from eq. (5.26). The total momentum exchange $\Delta \boldsymbol{p}$ at a given wall point is the sum over all contributions of populations which are bounced back at that point.

The wall stress tensor $\boldsymbol{\sigma}^\mathrm{w}$ and the momentum exchange $\Delta \boldsymbol{p}$ for an area element $\Delta \boldsymbol{A}$ per time step Δt are connected via

$$\frac{\Delta \boldsymbol{p}}{\Delta t} = \boldsymbol{\sigma}^\mathrm{w} \cdot \Delta \boldsymbol{A}. \tag{5.28}$$

The vector $\Delta \boldsymbol{A}$ is normal to the area element and points into the fluid. If $\Delta \boldsymbol{A}$ corresponds to one lattice area element, its magnitude is $|\Delta \boldsymbol{A}| = \Delta x^2$. For example—in lattice units—for a bottom wall in the xy-plane, $\Delta \boldsymbol{A} = \boldsymbol{e}_z$, the xz-component of the wall stress tensor obeys $\sigma_{xz}^\mathrm{w} = \Delta p_x$. For the corresponding top wall, $\Delta \boldsymbol{A} = -\boldsymbol{e}_z$, the WSS is $\sigma_{xz}^\mathrm{w} = -\Delta p_x$.

The walls shall be located at positions $z = \pm D/2$ (D is the distance between the walls), and the imposed shear stress component shall be σ_{xz} so that the resulting flow will have a velocity vector in x-direction. This is not a restriction, and it can be shown that the yz-component of the shear stress may be imposed independently, at least for the D3Q19 lattice which is considered throughout this thesis for 3D simulations. For a complex fluid, the apparent viscosity is not generally known a priori since it may depend on the internal structure of the fluid and its velocity field. This means that the bottom and top wall velocities, u_x^b and u_x^t, are not directly known if a shear stress is imposed. For the D3Q19 lattice, the only populations which are directly affected by an imposed value of σ_{xz} at the walls are f_7, f_8, f_{11}, and f_{12} since these are the only ones with non-vanishing x- and z-velocity components (fig. 5.1),

$$\begin{aligned}\boldsymbol{c}_7 &= (+1, 0, +1), & \boldsymbol{c}_8 &= (-1, 0, -1), \\ \boldsymbol{c}_{11} &= (-1, 0, +1), & \boldsymbol{c}_{12} &= (+1, 0, -1).\end{aligned} \tag{5.29}$$

For the yz-component of the stress tensor, only f_{15}, f_{16}, f_{17}, and f_{18} are required,

$$\begin{aligned}\boldsymbol{c}_{15} &= (0, -1, +1), & \boldsymbol{c}_{16} &= (0, +1, -1), \\ \boldsymbol{c}_{17} &= (0, -1, -1), & \boldsymbol{c}_{18} &= (0, +1, +1).\end{aligned} \tag{5.30}$$

In the following, the discussion is restricted to the top wall and the xz-component of the stress tensor. The analyses for the bottom wall and the yz-component are equivalent.

Taking two adjacent fluid nodes at $\boldsymbol{x}_\mathrm{l}$ and $\boldsymbol{x}_\mathrm{r}$ and the point $\boldsymbol{x}_\mathrm{w} = \boldsymbol{x}_\mathrm{l} + \frac{1}{2}\boldsymbol{c}_7 = \boldsymbol{x}_\mathrm{r} + \frac{1}{2}\boldsymbol{c}_{11}$ on the wall as shown in fig. 5.6, it follows from eq. (5.27) that the momentum exchange along the x-axis at the top wall is

$$\Delta p_x^\mathrm{t}(\boldsymbol{x}_\mathrm{w}, t+\tfrac{1}{2}) = \left(f_7^*(\boldsymbol{x}_\mathrm{l}, t) + f_8(\boldsymbol{x}_\mathrm{l}, t+1)\right) - \left(f_{11}^*(\boldsymbol{x}_\mathrm{r}, t) + f_{12}(\boldsymbol{x}_\mathrm{r}, t+1)\right) \tag{5.31}$$

[4]The momentum transfered from the wall to the fluid is the same with an additional minus sign. However, in this thesis, momentum transfer always means the momentum transfered from the fluid to the wall.

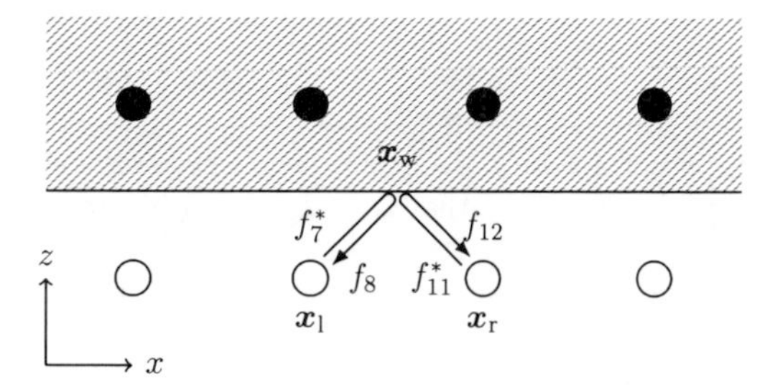

Fig. 5.6.: Momentum exchange at a solid wall in the lattice Boltzmann method. The physical wall (black line) separates the fluid region (white) and the fluid nodes (white circles) from the obstacle region (striped) and the obstacle nodes (black circles). The post-collision populations f_7^* and f_{11}^* propagate from $\boldsymbol{x}_\mathrm{l}$ and $\boldsymbol{x}_\mathrm{r}$ to the wall location $\boldsymbol{x}_\mathrm{w}$, respectively. At this point, both populations are bounced back and reach their starting points as post-propagation populations f_8 and f_{12}, respectively. During bounce-back, the momentum $\Delta p_{\mathrm{t}x}$, cf. eq. (5.31), is transferred from the populations to the wall.

and consequently

$$f_8(\boldsymbol{x}_\mathrm{l}, t+1) - f_{12}(\boldsymbol{x}_\mathrm{r}, t+1) = \Delta p_x^\mathrm{t}(\boldsymbol{x}_\mathrm{w}, t+\tfrac{1}{2}) + \big(f_7^*(\boldsymbol{x}_\mathrm{l}, t) - f_{11}^*(\boldsymbol{x}_\mathrm{r}, t)\big). \tag{5.32}$$

From eq. (5.26) it follows that

$$\begin{aligned} f_8(\boldsymbol{x}_\mathrm{l}, t+1) &= f_7^*(\boldsymbol{x}_\mathrm{l}, t) - 2w\rho(\boldsymbol{x}_\mathrm{l}, t)\frac{u_x^\mathrm{t}(\boldsymbol{x}_\mathrm{w}, t+\frac{1}{2})}{c_\mathrm{s}^2}, \\ f_{12}(\boldsymbol{x}_\mathrm{r}, t+1) &= f_{11}^*(\boldsymbol{x}_\mathrm{r}, t) + 2w\rho(\boldsymbol{x}_\mathrm{r}, t)\frac{u_x^\mathrm{t}(\boldsymbol{x}_\mathrm{w}, t+\frac{1}{2})}{c_\mathrm{s}^2}. \end{aligned} \tag{5.33}$$

Combining eq. (5.32) and eq. (5.33) yields

$$u_x^\mathrm{t}(\boldsymbol{x}_\mathrm{w}, t+\tfrac{1}{2}) = \frac{2\big(f_7^*(\boldsymbol{x}_\mathrm{l}, t) - f_{11}^*(\boldsymbol{x}_\mathrm{r}, t)\big) - \Delta p_x^\mathrm{t}(\boldsymbol{x}_\mathrm{w}, t+\frac{1}{2})}{2w\big(\rho(\boldsymbol{x}_\mathrm{l}, t) + \rho(\boldsymbol{x}_\mathrm{r}, t)\big)} c_\mathrm{s}^2. \tag{5.34}$$

This equation is an important intermediate result since it shows that the wall velocity at a given point is uniquely connected to the momentum exchange and the known post-collision values of the populations hitting the wall at that point. Assuming that f_7^* and f_{11}^* are known (which is usually the case), eq. (5.34) allows to compute either the momentum exchange, Δp_x^t, from the imposed wall velocity, u_x^t, or vice versa. While the former option is used in velocity BB BCs, the latter will be exploited in the remaining part of this section.

A non-rotating, stiff, and macroscopic wall must have the same velocity at each point. This means that u_x^t cannot be a function of position. Since, in general, f_7^* and f_{11}^* are functions of position along the wall, it cannot be expected that the momentum transfer Δp_x^t is homogeneously distributed over the wall. In other words, it is not possible to impose constant velocity and constant momentum transfer at the same time in a general flow situation. This is not a shortcoming of the numerical method. It is rather a physical effect, reflecting that the local wall stress depends on the local velocity profile of the nearby fluid. To this end, the problem of shear stress BCs has to be formulated in the following way: Given an imposed *average* momentum transfer, $\langle \Delta p_x^\mathrm{t} \rangle$, which homogeneous velocity u_x^t will result? Eq. (5.34) can be rewritten and brought in the form

$$u_x^\mathrm{t}(t+\tfrac{1}{2}) = \frac{2\,\langle f_7^*(t) - f_{11}^*(t)\rangle - \langle \Delta p_x^\mathrm{t}(t+\frac{1}{2})\rangle}{4w\,\langle \rho(t)\rangle} c_\mathrm{s}^2. \tag{5.35}$$

$\langle\cdot\rangle$ indicates an average along the entire top wall: The density and the populations are averaged in the fluid layer closest to the wall and the momentum transfer directly on the wall. For an imposed average momentum transfer, the velocity of the stiff wall follows from eq. (5.35). This velocity is then, as usual, used in eq. (5.26) to obtain the unknown populations f_8 and f_{12} at fluid nodes adjacent to the wall. The difference of this algorithm compared to the standard BB is that the wall velocity is not known a priori and that the resulting wall stress matches the imposed value. It has to be emphasized that this algorithm has to be used at each time step since the density, but more importantly the populations f_7^* and f_{11}^*, are functions of time in general. Just like for the standard velocity BB BC, the mass is conserved. However, the required rigidity of the wall leads to a non-locality for the shear stress BC: The surface averages of the involved populations and the density have to be computed.

The analogous equations for the bottom wall and the yz-cases read

$$u_x^{\mathrm{b}}(t+\tfrac{1}{2}) = -\frac{2\langle f_8^*(t) - f_{12}^*(t)\rangle + \left\langle \Delta p_x^{\mathrm{b}}(t+\tfrac{1}{2})\right\rangle}{4w\langle\rho(t)\rangle} c_{\mathrm{s}}^2, \tag{5.36}$$

$$u_y^{\mathrm{t}}(t+\tfrac{1}{2}) = \frac{(2\langle f_{18}^*(t) - f_{15}^*(t)\rangle - \left\langle \Delta p_y^{\mathrm{t}}(t+\tfrac{1}{2})\right\rangle}{4w\langle\rho(t)\rangle} c_{\mathrm{s}}^2, \tag{5.37}$$

$$u_y^{\mathrm{b}}(t+\tfrac{1}{2}) = -\frac{2\langle f_{17}^*(t) - f_{16}^*(t)\rangle + \left\langle \Delta p_y^{\mathrm{b}}(t+\tfrac{1}{2})\right\rangle}{4w\langle\rho(t)\rangle} c_{\mathrm{s}}^2. \tag{5.38}$$

Different signs in the equations for the bottom wall have to be noted. They can be easily understood when the case of simple shear flow is considered: In simple shear flow, the bottom and top wall shear stresses are identical, $\sigma_{xz}^{\mathrm{t}} = \sigma_{xz}^{\mathrm{b}}$, but the moment transfers have different signs, $\Delta p_x^{\mathrm{t}} = -\Delta p_x^{\mathrm{b}}$, cf. eq. (5.28). Furthermore, f_7 at the top wall plays the same role as f_8 at the bottom wall (the same also for f_{11} and f_{12}). The bottom wall moves in opposite direction as the top wall, $u_x^{\mathrm{b}}(t+\frac{1}{2}) = -u_x^{\mathrm{t}}(t+\frac{1}{2})$. This way, the signs in the equations can be verified.

It is an advantage that there are no body diagonals $(\pm1, \pm1, \pm1)$ in the D3Q19 lattice. It is easy to see that a population either participates in the σ_{xz} or the σ_{yz} BC algorithm, but not in both. This means that the xz- and the yz-BCs can be enforces independently. Furthermore, the xz-BC can be a stress BC, and the yz-BC can be a velocity BC at the same time. A benchmark test of this newly proposed and well-working BC is presented in section 9.5.

Concluding, the algorithm for the shear stress BC is outlined here again for the xz-component and the top wall on the D3Q19 lattice. The yz-component and the bottom wall work analogously and independently.

1. At each time step t, specify a desired average momentum exchange $\left\langle \Delta p_x^{\mathrm{t}}(t+\frac{1}{2})\right\rangle$ which is equivalent to the average xz-component of the top wall stress.
2. Compute the resulting top wall velocity $u_x^{\mathrm{t}}(t+\frac{1}{2})$ from eq. (5.35).
3. Perform the standard velocity BB BC algorithm in eq. (5.26) with this velocity in order to obtain the unknown populations $f_8(\boldsymbol{x}, t+1)$ and $f_{12}(\boldsymbol{x}, t+1)$ at the wall.

5.5. Efficiency and choice of simulation parameters in the lattice Boltzmann method

In the following, the significance of the simulation parameters in the LBM will be discussed. The relations between physical and lattice units are given in section 8.2.

The LBGK equation, eq. (5.3), has a free parameter, the dimensionless relaxation time τ. It can be selected in a given range to control the kinematic viscosity ν of the fluid (with unit $\mathrm{m^2\,s^{-1}}$) via eq. (5.7). The lattice Mach number is defined as

$$\mathrm{Ma} = \frac{\hat{u}}{c_{\mathrm{s}}} \tag{5.39}$$

where $\hat{u}$ is a characteristic (typically the maximum) velocity in the system, and c_s is the lattice speed of sound. For the plain NSE, the Reynolds number is the only relevant dimensionless parameter. It may be defined as

$$\mathrm{Re} = \frac{l\hat{u}}{\nu} \tag{5.40}$$

with a characteristic length l. This physical length is discretized by N lattice constants such that

$$l = N\Delta x. \tag{5.41}$$

Generally speaking: The larger N, the higher the spatial resolution. The smaller N, the faster the simulation.

It follows from eq. (5.7), eq. (5.39), eq. (5.40), and eq. (5.41) that

$$\frac{\mathrm{Ma}}{\mathrm{Re}} = \frac{\left(\tau - \frac{1}{2}\right) c_s}{N} \frac{\Delta t}{\Delta x}. \tag{5.42}$$

It should be noted again that for most of the available lattices, $c_s = \sqrt{1/3}\Delta x/\Delta t$ holds. Eq. (5.42) shows that only three of the four simulation parameters (τ, Ma, Re, N) can be chosen independently. Usually, the physical parameters for the simulation are known, in this case the Reynolds number. Thus, the researcher is left with the problem which two of the remaining three parameters (τ, Ma, N) to set first and how. This, generally, is a problem-specific task.

Instead of Ma and N, the time step Δt and the lattice constant Δx may be chosen as parameters. They are connected via

$$\nu = \frac{\tau - \frac{1}{2}}{3} \frac{\Delta x^2}{\Delta t}, \tag{5.43}$$

which can be obtained from eq. (5.7) and $c_s^2 = \Delta x/(3\Delta t)$.

The lattice Mach number must be sufficiently small in order to avoid truncation errors due to the Taylor expanded equilibrium, eq. (5.6), and compressibility errors caused by the equation of state of an ideal gas, eq. (5.10). For that reason, the lattice Mach number is usually kept below 0.1 or 0.2. The lattice resolution N can in principle be chosen freely. However, N must be sufficiently large in order to resolve regions of interesting physics, but it should be preferably small at the same time since the number of lattice nodes in the simulation grows like N^3 in 3D and with it the memory footprint and computing time. Additionally, on the one hand, the relaxation time cannot approach 0.5 arbitrarily closely. LB simulations are only stable for values of τ larger than $0.5 + \epsilon$ where ϵ is a small but finite and positive number which depends on the flow configuration and the lattice [163]. On the other hand, it is known that for τ larger than unity, the LBM becomes more and more inaccurate, especially in the presence of boundaries, as discussed in [150, 164, 165].

A common approach is to set τ and Δx or N first. From the known Reynolds number of the physical problem, the time step Δt (and thus the number of required time steps) and the lattice Mach number can then be inferred from eq. (5.43) and eq. (5.42), respectively. It must be emphasized that the lattice Mach number usually does not match the physical Mach number. This is not a problem in general as long as it is not too large (section 8.2). If, however, the lattice Mach number turns out to be too large (Ma > 0.2, say), either the lattice constant Δx or the relaxation time τ has to be decreased. This iteration has to be continued until all parameters have values which ensure (i) numerical stability, (ii) small compressibility artifacts, (iii) manageable lattice and (iv) time step sizes. It is easy to believe that the above considerations limit the applicability of the LBM as will be illustrated briefly by means of two examples.

For a high Mach number, $\mathrm{Ma} = 0.1$, a small viscosity, $\tau = 0.51$, and an intermediate resolution, $N = 100$, the Reynolds number as defined in eq. (5.40) is about 1700. This example illustrates that the LBM in its presented form is intrinsically not well suited for the simulation of flows with Reynolds numbers significantly larger than $\mathcal{O}(10^3)$ or $\mathcal{O}(10^4)$. Generally, high Reynolds number flows require large resolutions and a long computing time. Contrarily, extremely small Reynolds numbers ($< 10^{-3}$, say) are also difficult to reach. Assuming that a minimum resolution for an accurate flow description is required, $N = 15$, and that the relaxation parameter is taken as $\tau = 1$ to reduce numerical artifacts caused by higher values of the relaxation parameter, then the Reynolds number is about $\mathrm{Re} \approx 50\,\mathrm{Ma}$. In order to recover a small Reynolds number (e.g., < 0.1), the flow velocity u must be sufficiently small, which increases the number of time steps and the computational time proportionally to $1/u$. Therefore, the LBM reveals its strength in the intermediate Reynolds number regimes, usually between 0.1 and 100 [98].

The above discussion is made even more complex when the scaling of the error terms in the LBM is considered (appx. B.1.2). It can be shown that the error in the LBM is proportional to $\mathcal{O}(\Delta x^2) + \mathcal{O}(\mathrm{Ma}^2)$ (e.g., [126, 165]). An error analysis in the LBM with respect to the choice of the simulation parameters can be found in [164, 165]. One important result is that a relaxation parameter τ somewhere between 0.8 and 1.0 minimizes the associated numerical errors. A discussion of the choice of simulation parameters is also provided in [166].

6. Fluid-structure interaction: the immersed boundary method

The immersed boundary method (IBM) is presented in this chapter. Its important task is to provide the bidirectional coupling between the fluid motion and the membrane dynamics. The basic idea of this coupling is that the membranes move along with the ambient fluid (no-slip condition at the interface) and that any force acting on the membranes also acts on the fluid and vice versa (Newton's third law). After an overview in section 6.1, the governing equations are motivated in section 6.2. The discretization of the IBM equations is discussed in section 6.3, followed by an alternative motivation of these equations based on statistical physics in section 6.4.

6.1. Overview of the immersed boundary method

The IBM has been introduced by Peskin in the 1970s to simulate the blood flow around heart valves [78, 167, 168]. The purpose of the IBM is the computational modeling of fluid-structure interactions. The mathematical basis consists of two coordinate systems, an Eulerian and a Lagrangian system. The Eulerian variables are defined on a fixed Cartesian mesh while the Lagrangian quantities live on a curvilinear or unstructured mesh which may move on top of the Eulerian mesh. The Eulerian mesh is used to solve the Navier-Stokes equations (NSE) while the Lagrangian system captures the immersed structures (e.g., membranes) in the fluid. In general, the two meshes are not conform (fig. 6.1), which raises the need of interpolations when information is transferred from one mesh to the other. The IBM is a front-tracking coupling method, i.e., the interface location is explicitly known. Unlike the bounce-back scheme (section 5.4), the IBM acts via body forces on the fluid in order to enforce the boundary conditions (BCs) resulting from the presence of the structures immersed in the fluid.

One of the basic assumptions of the IBM is the validity of the no-slip condition, i.e., each immersed structure element moves with the same velocity as the ambient fluid. Conversely, the structure exerts a force on the nearby fluid which enters the NSE as an external forcing term. This force mimics the momentum exchange of the fluid at the structure surface and can also be interpreted as the force obtained from the constitutive model of the elastic immersed material (chap. 7). The forces acting on the fluid are originally computed in the Lagrangian frame of the structure. Thus, the forces have to be spread to the Eulerian mesh in order to solve the NSE. The resulting fluid velocity has to be interpolated back to the Lagrangian mesh for the update of the structure element positions. The underlying interpolation functions (also called interpolation stencils) for the force spreading and the velocity interpolation have to be defined in a consistent way. The algorithm and discretizations are given in sections 6.2 and 6.3.

The IBM offers a number of advantages. First, it can be combined with any Navier-Stokes solver which supports external forcing (e.g., the LBM). Second, the constitutive behavior of the immersed elastic structures is not restricted by the IBM. In that sense, the IBM is a pure coupling method obeying the no-slip condition at the fluid-structure interface. A further advantage is that there are no additional, unphysical parameters in the IBM which have to be tuned or optimized. The implementation of the IBM is comparably simple, and its numerical overhead is small. Particularly with regard to the simulation of suspensions of deformable particles at high

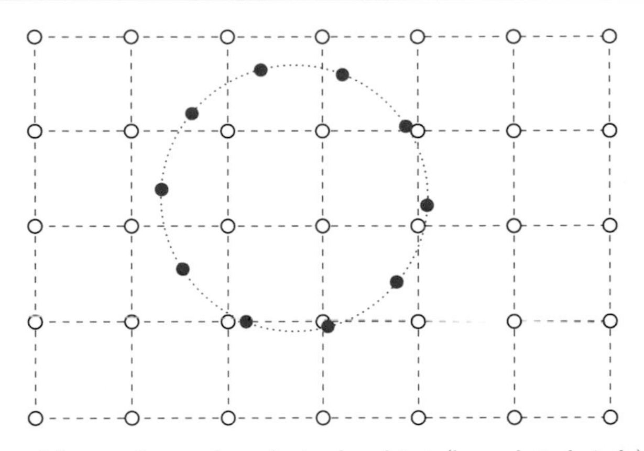

Fig. 6.1.: Eulerian and Lagrangian meshes. A circular object (large dotted circle) of radius $1.4\Delta x$ is described by an ensemble of Lagrangian points (red circles) on the background of the regular Eulerian grid (dashed lines and white circles). Generally, the Lagrangian mesh is not conform with the Eulerian grid.

volume fractions, an important advantage of the IBM is that arbitrarily complex fluid-structure interfaces can be modeled. Numerical problems related to the simulation of dense suspensions of deformable particles are discussed separately in section 8.6.

The IBM has been used before in connection with the LBM or other Navier-Stokes solvers in order to simulate suspensions of soft particles [77, 80, 89, 169, 170]. Apart from the application being in the focus of this work (simulation of deformable particles), the IBM is regularly employed to model rigid objects immersed in a fluid. Peskin [78] provides a rich collection of applications and extensions of the IBM such as improving the time-stepping scheme [171], the volume conservation [172], sharpening its interface [173], implementing local grid refinement [174], and parallelizing the IBM [175]. Some applications include Peskin's original work about fluid dynamics of heart valves [167], simulation of particle suspensions [176], platelet aggregation during blood clotting [177], flow in elastic blood vessels [178], simulation of biofilms [179], and flow past a cylinder [180], to name only a few. Additional applications of the IBM are reviewed in a more recent article by Mittal and Iaccarino [181].

6.2. Governing equations of the immersed boundary method

A thorough mathematical derivation of the IBM formalism has been provided by Peskin [78]. It shall not be repeated here. In the following, solely the governing equations and some remarks are collected for a special case of the IBM. It is assumed that

- the immersed structure is a 2D membrane immersed in 3D space and that
- the density of the membrane equals the density of the ambient fluid.

Both assumptions are reasonable when red blood cells (RBCs) are simulated (chap. 4). However, in general, the IBM may also be applied to situations where the above assumptions cannot be made. The corresponding generalized equations can be found in [78].

Let $\boldsymbol{X}$ be the coordinate of a fixed point in the Eulerian frame and $\boldsymbol{x}(r, s, t)$ the position of a marker point comoving with the Lagrangian mesh. (r, s) are two-dimensional curvilinear coordinates for the membrane. The exact form of the curvilinear coordinate system is not important since it will not appear in the discretized equations at the end. Still, it is instructive to use (r, s) in order to better understand the IBM formalism. The governing equations for the fluid-membrane coupling read [78]

$$\boldsymbol{f}(\boldsymbol{X},t) = \int \mathrm{d}r\,\mathrm{d}s\,\tilde{\boldsymbol{f}}(r,s,t)\delta(\boldsymbol{X}-\boldsymbol{x}(r,s,t)), \tag{6.1}$$

$$\dot{\boldsymbol{x}}(r,s,t) = \int \mathrm{d}^3X\,\boldsymbol{u}(\boldsymbol{X},t)\delta(\boldsymbol{X}-\boldsymbol{x}(r,s,t)). \tag{6.2}$$

Here, $\delta(\boldsymbol{X}-\boldsymbol{x}(r,s,t))$ is the three-dimensional Dirac delta distribution. $\boldsymbol{u}(\boldsymbol{X},t)$ is the velocity of the fluid at coordinate $\boldsymbol{X}$ at time t, and $\dot{\boldsymbol{x}}(r,s,t)$ is the velocity of the Lagrangian marker point $\boldsymbol{x}(r,s,t)$. $\boldsymbol{f}(\boldsymbol{X},t)$ is the force density (force per volume) acting on the fluid at coordinate $\boldsymbol{X}$ and time t. The force density (force per area) in the Lagrangian system at position $\boldsymbol{x}(r,s,t)$ is denoted by $\tilde{\boldsymbol{f}}(r,s,t)$. Eq. (6.2) resembles the no-slip condition at the membrane surface.

It has to be noted that eq. (6.1) and eq. (6.2) behave differently, even though they have the same interaction function $\delta(\boldsymbol{X}-\boldsymbol{x}(r,s,t))$. For a 2D membrane, force densities are area densities. Thus, on the one hand, the force density $\boldsymbol{f}(\boldsymbol{X},t)$ on the left-hand-side of eq. (6.1) is singular like a one-dimensional delta function since the integral is only 2D. On the other hand, the velocities $\dot{\boldsymbol{x}}(r,s,t)$ and $\boldsymbol{u}(\boldsymbol{X},t)$ in eq. (6.2) are both finite. The transformation in eq. (6.1) is called spreading, and the transformation in eq. (6.2) is called interpolation [78].

For the simulations in the present work, eq. (6.1) and eq. (6.2) form the interaction equations between the membranes and the ambient fluid. In the next step, it has to be discussed how the mathematical relations can be discretized in order to use them in numerical simulations.

6.3. Discretization of the immersed boundary method

The discretization of the IBM equations, eq. (6.1) and eq. (6.2), is necessary to implement the model into a numerical scheme. Especially, a reasonable discretized delta function has to be found. In the following, the spatial discretization scheme will be discussed. The time discretization is shortly presented at the end of this section. Omitted intermediate steps and further comments can be found in [78].

In their discretized forms, the spreading and interpolation equations, eq. (6.1) and eq. (6.2), read

$$\boldsymbol{f}(\boldsymbol{X},t) = \sum_{r,s} \tilde{\boldsymbol{f}}(r,s,t)\delta_\Delta(\boldsymbol{X}-\boldsymbol{x}(r,s,t))\,\Delta r\,\Delta s, \tag{6.3}$$

$$\dot{\boldsymbol{x}}(r,s,t+\Delta t) = \sum_{\boldsymbol{X}} \boldsymbol{u}(\boldsymbol{X},t+\Delta t)\delta_\Delta(\boldsymbol{X}-\boldsymbol{x}(r,s,t))\,\Delta x^3 \tag{6.4}$$

where the integration is replaced by a discrete sum and $\delta_\Delta(\boldsymbol{X}-\boldsymbol{x}(r,s,t))$ is the discretized delta function. The time increment is denoted Δt. Δx, Δr, and Δs are the Eulerian lattice constant and the sizes of the Lagrangian membrane elements, respectively. The velocity interpolation and force spreading are illustrated in fig. 6.2.

It has to be stressed again that $\tilde{\boldsymbol{f}}(r,s,t)$ is the force density of the membrane (force per area), whereas $\boldsymbol{F}(r,s,t) = \tilde{\boldsymbol{f}}(r,s,t)\,\Delta r\,\Delta s$ is the force acting on the membrane area defined by $(\Delta r, \Delta s)$. In the discretized formulation, there is a given number of points (nodes) defining the membrane surface. Each of these points can be addressed either by its coordinates (r,s) or by a node index i. A curvilinear coordinate system is not necessarily required, and the mesh can be unstructured (which will be the case throughout this thesis, cf. section 8.3). Replacing (r,s) by the node index i and setting $\Delta x = \Delta t = 1$ in the following, the discretized IBM equations read

$$\boldsymbol{f}(\boldsymbol{X},t) = \sum_{i} \boldsymbol{F}_i(t)\delta_\Delta(\boldsymbol{X}-\boldsymbol{x}_i(t)), \tag{6.5}$$

$$\dot{\boldsymbol{x}}_i(t+1) = \sum_{\boldsymbol{X}} \boldsymbol{u}(\boldsymbol{X},t+1)\delta_\Delta(\boldsymbol{X}-\boldsymbol{x}_i(t)). \tag{6.6}$$

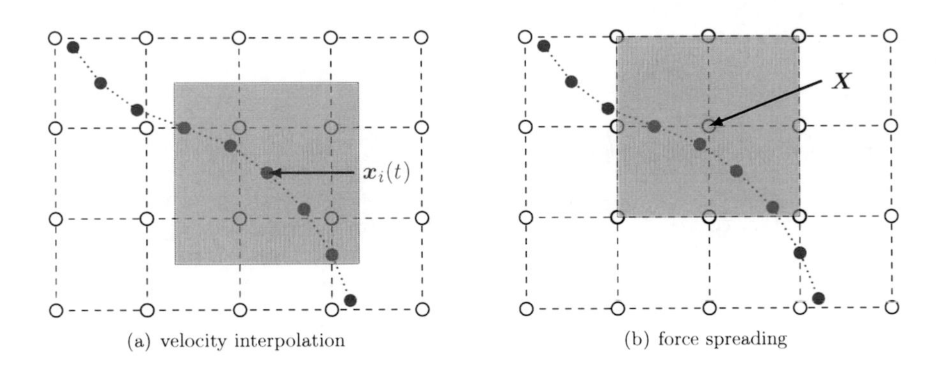

Fig. 6.2.: Velocity interpolation and force spreading in the immersed boundary method. A membrane patch is denoted by the curved dotted line. During (a) velocity interpolation, each membrane node (red circles) at position $\boldsymbol{x}_i(t)$ collects velocity information of all lattice nodes within a finite range (square box). During (b) force spreading, each lattice node (white circles) at fixed position $\boldsymbol{X}$ collects force information of all membrane nodes within a finite range (square box). The weights of the interpolation/spreading contributions are given by the value of the discrete delta function, e.g., eq. (6.11).

In this simplified picture, $\boldsymbol{F}_i(t)$ denotes the total force acting on node i which is located at position $\boldsymbol{x}_i(t)$ and has velocity $\dot{\boldsymbol{x}}_i(t)$.

The coupled fluid-membrane system is solved iteratively. For that reason, the positions at the old time step, $\boldsymbol{x}_i(t)$, are used in eq. (6.6) to update the membrane velocity and obtain its value at the next time step, $\dot{\boldsymbol{x}}_i(t+\Delta t)$. These algorithmic details are further elaborated on in section 8.1.

Obviously, a discussion of the discretized delta function $\delta_\Delta(\boldsymbol{X}-\boldsymbol{x}_i(t))$ is still missing in order to complete the discretization. According to Peskin [78], the discretized delta function has to obey a series of restrictions and properties, e.g., the force and the torque should be the same when evaluated in the Eulerian and the Lagrangian systems. Additionally, the discretized delta function should be continuous, which assures that there are neither jumps in the velocity nor in the force when the membrane points move between Eulerian lattice nodes. A complete list of those restrictions and their mathematical significance can be found in [78]. In order to increase computational efficiency, the discretized delta function should have a compact support, i.e., for each Lagrangian mesh point, only the Eulerian fluid points within a finite range should be considered and vice versa. The smallest possible support for realizing all of Peskin's postulates is four Eulerian grid points along each spatial dimension. It can be shown that the same discretized delta function has to be used for spreading and interpolation [78].

One of the major assumptions is that the discretized delta function can be factorized,

$$\delta_\Delta(\boldsymbol{x}) = \phi(x)\phi(y)\phi(z). \tag{6.7}$$

This ansatz is not essential, but the computations become simpler, and the cubic lattice structure is taken into account.

It is possible to find various discretized delta functions which have different interpolation ranges. The so-called 4-point stencil reads

$$\phi_4(x) = \begin{cases} \frac{1}{8}\left(3-2|x|+\sqrt{1+4|x|-4x^2}\right) & \text{for} \quad 0 \le |x| \le 1, \\ \frac{1}{8}\left(5-2|x|-\sqrt{-7+12|x|-4x^2}\right) & \text{for} \quad 1 \le |x| \le 2, \\ 0 & \text{for} \quad 2 \le |x|. \end{cases} \tag{6.8}$$

This discretization fulfills all restrictions which are stated by Peskin [78]. The interpolation function

$$\phi_4^c(x) = \begin{cases} \frac{1}{4}(1+\cos(\frac{\pi x}{2})) & \text{for} \quad 0 \le |x| \le 2, \\ 0 & \text{for} \quad 2 \le |x| \end{cases} \tag{6.9}$$

which is an excellent approximation of eq. (6.8) is regularly used in the literature instead.

One can construct an interaction function with a support of three lattice nodes,

$$\phi_3(x) = \begin{cases} \frac{1}{3}(1+\sqrt{1-3x^2}) & \text{for} \quad 0 \le |x| \le \frac{1}{2}, \\ \frac{1}{6}(5-3|x|-\sqrt{-2+6|x|-3x^2}) & \text{for} \quad \frac{1}{2} \le |x| \le \frac{3}{2}, \\ 0 & \text{for} \quad \frac{3}{2} \le |x|. \end{cases} \tag{6.10}$$

Similar to $\phi_4(x)$ and $\phi_4^c(x)$, $\phi_3(x)$ is symmetric, $\phi(-x) = \phi(x)$, and it has a continuous first derivative. These two properties have not been claimed, but come in handy. Peskin [78] and Dünweg and Ladd [182] state that Navier-Stokes solvers depending on a central difference scheme cannot use $\phi_3(x)$ since the number of support points is odd. The LBM, however, is not concerned with this restriction. Obviously, there are some advantages of $\phi_3(x)$ over $\phi_4(x)$. First, the envelope volume is decreased from 64 to 27 grid points in 3D reducing the computational overhead. Second, the membrane interface width is decreased. It is reported by Dünweg and Ladd [182] that $\phi_3(x)$ results in hydrodynamics which is nearly as accurate as that for $\phi_4(x)$.

If maximum efficiency is required, it is also possible to use a two-point linear interaction function with a support of two lattice nodes along each axis,

$$\phi_2(x) = \begin{cases} 1-|x| & \text{for} \quad 0 \le |x| \le 1, \\ 0 & \text{for} \quad 1 \le |x|. \end{cases} \tag{6.11}$$

This way, the cubic lattice structure becomes more visible, i.e., the translational symmetry is violated more strongly than for ϕ_3 or ϕ_4 [78, 182]. Obviously, ϕ_2 does not have a continuous derivative, but only eight lattice nodes have to be considered for spreading and interpolation. The shapes of the three discretized delta functions ϕ_2, ϕ_3, and ϕ_4 are shown in fig. 6.3.

In the present work, for reasons of numerical efficiency and for reducing the numerical membrane interface width, usually ϕ_2 is employed (as also in, e.g., [183]). It should be noted that even different discretized delta functions may be used [78, 184, 185].

When the IBM is combined with the LBM, the explicit Euler method is usually employed for the time discretization (e.g., [77]),

$$\boldsymbol{x}_i(t+\Delta t) = \boldsymbol{x}_i(t) + \dot{\boldsymbol{x}}_i(t+\Delta t)\Delta t. \tag{6.12}$$

However, there exist different time integration schemes, [186, 187].

6.4. Connection between the immersed boundary method and viscous coupling

Eq. (6.1) and eq. (6.2) can also be derived from a more general method. Dünweg and Ladd [182] use a fluctuating LBM in connection with particles dissipatively coupled to the fluid. These particles experience a drag force if their velocity differs from the ambient fluid velocity. Dropping the time from the following equations for simplicity, the drag force acting on particle i is

$$\boldsymbol{F}_i^{\mathrm{d}} = -\Gamma_i \left(\frac{\boldsymbol{p}_i}{m_i} - \boldsymbol{u}(\boldsymbol{x}_i) \right) \tag{6.13}$$

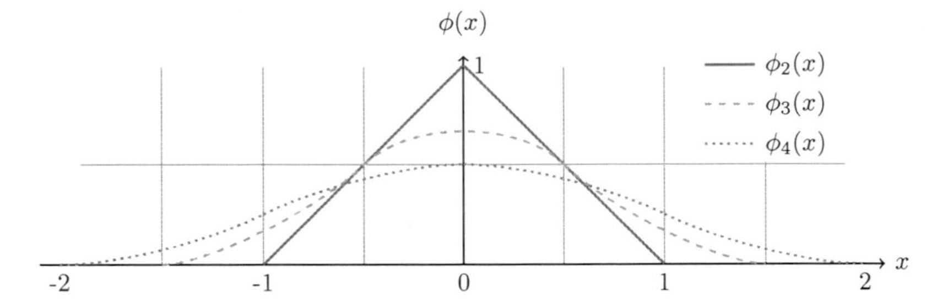

Fig. 6.3.: Discrete delta functions for the immersed boundary method (IBM). The 2-point (solid), 3-point (dashed), and 4-point (dotted) discrete delta functions for the IBM are shown.

where Γ_i is the drag coefficient for the particle, $\boldsymbol{p}_i$ is its momentum, m_i is its mass, and $\boldsymbol{u}(\boldsymbol{x}_i)$ is the fluid velocity at the position of the particle. The equations of motion for the particles are

$$\frac{\mathrm{d}}{\mathrm{d}t}\boldsymbol{x}_i = \frac{1}{m_i}\boldsymbol{p}_i, \tag{6.14}$$

$$\frac{\mathrm{d}}{\mathrm{d}t}\boldsymbol{p}_i = \boldsymbol{F}_i^{\mathrm{c}} + \boldsymbol{F}_i^{\mathrm{d}} + \boldsymbol{F}_i^{\mathrm{f}}. \tag{6.15}$$

Here, $\boldsymbol{F}_i^{\mathrm{c}}$ is the conservative force acting on the particle (e.g., due to an external potential or particle interactions), and $\boldsymbol{F}_i^{\mathrm{f}}$ is a Langevin noise for the particle. According to [182], the force density applied to the fluid at Eulerian coordinate $\boldsymbol{X}$ is computed numerically from

$$\boldsymbol{f}(\boldsymbol{X}) = -\sum_i \left(\boldsymbol{F}_i^{\mathrm{d}}(\boldsymbol{x}_i) + \boldsymbol{F}_i^{\mathrm{f}}(\boldsymbol{x}_i)\right) \delta_\Delta(\boldsymbol{X} - \boldsymbol{x}_i) \tag{6.16}$$

with the same discretized delta functions as in section 6.3. It has been shown by Dünweg and Ladd [182] that the fluctuation-dissipation theorem holds for this coupled system. In the following, it will be inferred that the IBM is formally a special case of the viscous coupling. Consequently, the fluctuation-dissipation theorem should also hold for the IBM [188].

Eq. (6.14) and eq. (6.15) can be combined to give

$$m_i \frac{\mathrm{d}^2}{\mathrm{d}t^2}\boldsymbol{x}_i = \boldsymbol{F}_i^{\mathrm{c}} + \boldsymbol{F}_i^{\mathrm{d}} + \boldsymbol{F}_i^{\mathrm{f}} \tag{6.17}$$

which becomes

$$\boldsymbol{F}_i^{\mathrm{d}} + \boldsymbol{F}_i^{\mathrm{f}} = -\boldsymbol{F}_i^{\mathrm{c}} \tag{6.18}$$

in the over-damped, i.e., massless limit ($m_i \to 0$). Combining eq. (6.18) with eq. (6.16) directly results in the IBM force spreading equation, eq. (6.5), if the conservative force is identified as the membrane force. This finding also justifies that, in the present model, the elastic (conservative) membrane force is used in eq. (6.5) to drive the fluid (chap. 7).

In the last step, keeping $m_i \to 0$, eq. (6.13), eq. (6.14), and eq. (6.18) are combined, which yields

$$\dot{\boldsymbol{x}}_i = \boldsymbol{u}(\boldsymbol{x}_i) + \frac{1}{\Gamma_i}(\boldsymbol{F}_i^{\mathrm{c}} + \boldsymbol{F}_i^{\mathrm{f}}). \tag{6.19}$$

In the high friction limit ($\Gamma_i \to \infty$), the no-slip condition is recovered and with it the IBM velocity interpolation, eq. (6.6).

The parameters m_i and Γ_i are purely numerical without any physical significance. In this sense, the IBM is more natural since it does not introduce additional parameters. However, the time steps for the IBM and the fluid solver are required to be identical. This is not the case in the approach followed by Dünweg and Ladd [182] where the molecular dynamics time step for the particles can be chosen much smaller then the hydrodynamic time step.

7. Membrane model and energetics

The computational model for the red blood cell (RBC) membrane is presented in this chapter. Based on the discussions in section 4.2, it is assumed that an undeformed RBC is stress-free and that this equilibrium shape does not change in time. The membrane model is formulated in such a way that any deviation from the equilibrium shape increases the membrane energy and response forces are induced which drive the membrane shape towards its equilibrium. The equilibrium shape itself can be chosen arbitrarily if other deformable particles shall be studied. In particular, the RBC membrane may be replaced by spherical, ellipsoidal, or more complex shapes.

Four relevant energy contributions for a RBC can be identified: Local in-plane forces are caused by the resistance to shear and dilation (section 7.1). The bilayer character of the membrane is taken into account by defining a local bending energy giving rise to forces normal to the membrane (section 7.2). Additionally, the total surface and volume of a RBC are virtually constant. This can be handled by introducing appropriate surface and volume energies as in sections 7.3 and 7.4, respectively. Forces acting between pairs of membranes or membranes and walls are introduced in section 8.7 and are not part of the membrane model itself.

It is assumed that the above-mentioned energy contributions are independent of each other, which is a common idealization in simulations [31]. For that reason, the energy contributions may also be neglected individually or amended by additional contributions if another type of membrane is considered. This renders the model far more general than it may seem on the first glance. The considerations in this chapter can thus be applied to other types of elastic membranes, not only to RBCs. It has to be noted as well that a volume deviation does not generally imply a surface deviation and vice versa. Both can occur independently. The same holds for the strain and bending contributions: A membrane can be bent without being sheared and sheared without being bent.

Deriving the membrane forces from appropriate energies guarantees momentum and angular momentum conservation. This is of high relevance because deformed particles are not expected to translate or rotate spontaneously. For that reason, no ad-hoc membrane forces are considered which may violate either the momentum or the angular momentum conservation.

7.1. Membrane strain and area dilation energetics

Gradient displacement tensor

The RBC membrane is considered to be a hyperelastic material, i.e., the stress-strain relationship derives from a strain energy area density ϵ^{S}. The strain energy of the membrane is the surface integral $E_{\mathrm{S}} = \oint \mathrm{d}A\, \epsilon^{\mathrm{S}}$.

Assuming an isotropic and homogeneous material [31], the energy density ϵ^{S} cannot depend on the orientation or location of the membrane. Instead, the energy is locally stored in a shear deformation and dilation of the membrane via the displacement gradient tensor $\boldsymbol{D}$. This tensor describes the local deformation state (strain and dilation) of the membrane.

In order to understand the concept of the deformation gradient, it is instructive to consider a

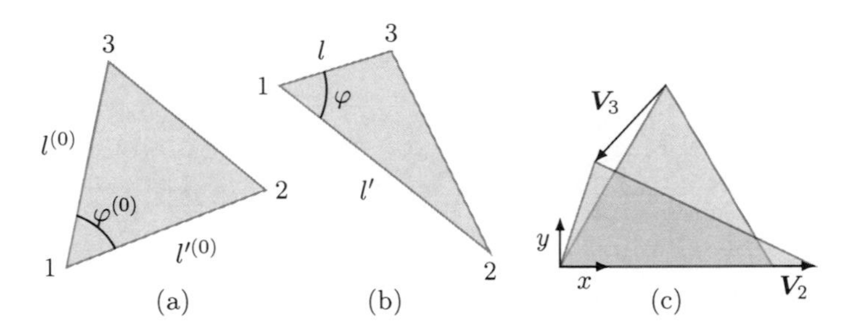

Fig. 7.1.: Deformation of a membrane face element. Each face is made up of three nodes (1, 2, 3) which define the edges l (between nodes 1 and 3) and l' (between nodes 1 and 2) and the angle φ between these two edges. (a) The equilibrium face (defined by $l^{(0)}$, $l'^{(0)}$, and $\varphi^{(0)}$), (b) its deformed shape (accordingly defined by l, l', and φ), and (c) both transformed to the same xy-plane are shown. The displacement vector $\boldsymbol{V}_1$ is identically zero, and the other two are shown in (c). The deformation state (λ_1, λ_2) of the face is then uniquely defined.

finite and flat triangular patch of the membrane as shown in fig. 7.1. Indeed, the membrane is numerically described by a number N_f of flat triangular face elements (finite elements, see below). The corresponding mesh generation procedure is described in section 8.3. In fig. 7.1, an initially undeformed triangular patch of the membrane with zero strain energy (a) is deformed (b). Due to the claim that rotations and translations do not change the energy, it is possible—without loss of generality—to align the undeformed and the deformed patches as shown in (c). By definition, the first node of the triangular patch is not shifted (no displacement), whereas the other two nodes generally have non-zero displacements $\boldsymbol{V}_2$ and $\boldsymbol{V}_3$. The displacement gradient tensor describes how the continuous displacement $\boldsymbol{V}$ varies over the face element since $\boldsymbol{V}$ is obviously not constant. Else, it could be chosen to be zero everywhere, and the face element would not be deformed at all. For membranes of negligible thickness, it is sufficient to tread the 2D case. Here, the displacement gradient tensor is defined as

$$\boldsymbol{D} = \begin{pmatrix} D_{xx} & D_{xy} \\ D_{yx} & D_{yy} \end{pmatrix} := \begin{pmatrix} 1 & 0 \\ 0 & 1 \end{pmatrix} + \begin{pmatrix} \partial_x V_x & \partial_y V_x \\ \partial_x V_y & \partial_y V_y \end{pmatrix}. \tag{7.1}$$

In the present model (see below), it is assumed that the displacement *gradient* is constant over a given face element, i.e., the displacement varies linearly over the face. This is the spirit of the linear finite element method (FEM).

Obviously, the energy of a face can only depend on the invariants of the tensor $\boldsymbol{D}$. These (in 2D two) invariants are the eigenvalues λ_1 and λ_2 of the tensor $\boldsymbol{D}$, and they are also called the principal in-plane stretch ratios. They do not depend on the origin or rotation of the coordinate system and can thus be considered as the desired physical deformation parameters. Equivalently, the so-called strain invariants $I_1 = \lambda_1^2 + \lambda_2^2 - 2$ and $I_2 = \lambda_1^2\lambda_2^2 - 1$ which describe the strain and dilation state of the membrane, respectively, can be used. For a general 3D elastic material, the tensor $\boldsymbol{D}$ is a 3×3-tensor and a third eigenvalue, λ_3, must be taken into account. More detailed explanations and the underlying theory can be found in textbooks about elasticity, e.g., [189].

Constitutive model

The energy of the homogeneous and isotropic membrane can only depend on the invariants λ_1 and λ_2 or I_1 and I_2. Therefore, an appropriate function $\epsilon^{\mathrm{S}}(I_1, I_2)$ for the energy density has to be found which describes the membrane material. This function is called the constitutive model, and its form is not fixed by the theory of elasticity. Instead, it must be chosen in such a way that it describes the stress-strain behavior of the material as closely as possible.

Deformations of biological cells can be large, thus, a linear stress-strain approximation is not justified in general. Skalak et al. [190] have suggested an energy model which is able to reproduce experimental data of RBCs at both small and large strains,

$$\epsilon^{\mathrm{S}} = \frac{\kappa_{\mathrm{S}}}{12}\left(I_1^2 + 2I_1 - 2I_2\right) + \frac{\kappa_\alpha}{12} I_2^2. \tag{7.2}$$

The surface elastic shear modulus κ_{S} and area dilation modulus κ_α control the strength of the membrane response to deformation (shear and dilation). Another, commonly used model is the neo-Hookean law which is equivalent to the zero-thickness shell membrane proposed by Ramanujan and Pozrikidis [191] for small deformations. Another constitutive model has been proposed by Navot [192]. More information about these constitutive laws can be found in the literature (e.g., [169, 193]). The investigation of quantitative and qualitative differences of these models is not in the scope of this thesis. For this reason, if not otherwise stated, the Skalak membrane model, eq. (7.2), will be employed.

Numerical procedure

In the following, the numerical evaluation of the deformation gradient and the corresponding membrane forces is described. On the one hand, within the present model, the membrane forces are treated as being concentrated at the corners of the faces (nodes or vertices). On the other hand, the deformation state (λ_1 and λ_2 or equivalently I_1 and I_2) is a property of the faces.

The first step in the computation of the strains $\lambda_{1,2}$ of a given face element is the identification of the displacements $\boldsymbol{V}_k$ ($k = 1, 2, 3$) of the nodes as shown in fig. 7.1. The deformed and undeformed faces are transformed to a common plane (here: xy-plane) in such a way that the edges $l'^{(0)}$ and l' are aligned (cf. fig. 7.1 for the node and edge conventions). It should be noted that both the undeformed and deformed faces are treated as flat triangles. The basic assumption is that the two-dimensional displacement gradient tensor $\boldsymbol{D}$, eq. (7.1), is spatially constant over the entire face. This can be realized by introducing a linear shape function $N_k(x, y) = a_k x + b_k y + c_k$ ($k = 1, 2, 3$) for each node in the face. The coefficients are found by letting $N_k(x_j, y_j) = \delta_{kj}$ ($k, j = 1, 2, 3$), i.e., each shape function N_k is unity at the location of the corresponding node k, but zero at the two nodes other than k. The linear displacement field of the face can then be written as

$$\boldsymbol{V}(x, y) = N_1\boldsymbol{V}_1 + N_2\boldsymbol{V}_2 + N_3\boldsymbol{V}_3, \tag{7.3}$$

and the displacement gradient tensor $\boldsymbol{D}$ is computed from eq. (7.1). Due to the linearity of the shape functions, the components $D_{\alpha\beta}$ do not depend on x or y but on the shape function coefficients a_k and b_k which are uniquely fixed by the shape of the undeformed element, i.e., a_k and b_k are constant in time for each node in the face. It should be noted that also non-linear shape functions may be used [191, 194]. However, due to numerical efficiency, the model is restricted to a linear approach. It is shown in appx. C.1.1 that the displacement gradient tensor then has the form $\boldsymbol{D} = \left(\begin{smallmatrix} a & b \\ 0 & c \end{smallmatrix}\right)$ with [31]

$$a = \frac{l'}{l'^{(0)}}, \quad b = \frac{1}{\sin\varphi^{(0)}}\left(\frac{l}{l^{(0)}}\cos\varphi - \frac{l'}{l'^{(0)}}\cos\varphi^{(0)}\right), \quad c = \frac{l}{l^{(0)}}\frac{\sin\varphi}{\sin\varphi^{(0)}}. \tag{7.4}$$

All symbols are defined in fig. 7.1.

The current deformation of a face is evaluated from the equations

$$\lambda_1^2\lambda_2^2 = a^2c^2, \tag{7.5}$$

$$\lambda_1^2 + \lambda_2^2 = a^2 + b^2 + c^2 \tag{7.6}$$

since $\lambda_{1,2}$ are the eigenvalues of the tensor $\boldsymbol{D}$ and $\lambda_1^2 + \lambda_2^2 = \mathrm{tr}(\boldsymbol{D}^\mathrm{T}\boldsymbol{D})$ and $\lambda_1^2\lambda_2^2 = \det(\boldsymbol{D}^\mathrm{T}\boldsymbol{D})$ where the superscript T denotes the matrix transpose. The product $\boldsymbol{D}^\mathrm{T}\boldsymbol{D}$ is rotationally invariant. More details are also provided in [31, 195, 196].

Numerically, the strain energy E_S is computed from the area energy density ϵ^S, eq. (7.2), and the reference (undeformed) area $A^{(0)}$ of the membrane faces, $E_\mathrm{S} = \sum_i A_i^{(0)}\epsilon_i^\mathrm{S}$, where the sum runs over all faces of the membrane [31, 195, 196]. The strain force of node i at position $\boldsymbol{x}_i$ caused by the deformation can be computed from the principle of virtual work [195, 196],

$$\boldsymbol{F}_i^\mathrm{S} = -\frac{\partial E_\mathrm{S}(\{\boldsymbol{x}_i\})}{\partial \boldsymbol{x}_i}. \tag{7.7}$$

The functional form of the forces is derived and presented in appx. C.1.2. In the undeformed state of the membrane ($I_1 = I_2 = 0$ for each face element), the total energy E_S and all forces $\boldsymbol{F}_i^\mathrm{S}$ are zero.

7.2. Membrane bending energetics

The bending energy of a RBC may be written in the form [31, 40]

$$E_\mathrm{B} = \frac{\kappa_\mathrm{B}}{2}\oint \mathrm{d}A\,\left(H - H^{(0)}\right)^2. \tag{7.8}$$

Here, H is the trace of the surface curvature tensor, and $H^{(0)}$ is the spontaneous curvature. The energy scale is given by the bending modulus κ_B. The energy term in eq. (7.8) is also denoted the *Helfrich* term, named after Helfrich who has first stipulated an energy term containing the spontaneous curvature of the membrane [197]. The investigation of the bending properties of RBCs goes back to Rand and Burton [198], Canham [199], and Evans [36] to name only a few investigators.

In principle, there are two additional terms contributing to the bending energy in eq. (7.8) [31]. The first is the Gaussian term $\kappa_\mathrm{G}\oint \mathrm{d}A\,K$ where K is the determinant of the curvature tensor and κ_G is the Gaussian bending modulus. One can show that this integral is constant as long as the topology of the membrane does not change, i.e., as long as a closed membrane remains closed [31, 40, 200, 201]. This is a consequence of the Gauss-Bennet theorem which states that $\oint \mathrm{d}A\,K = 4\pi$ for a simply connected surface with the topology of a sphere. Since the present membrane model does not allow rupture or topology change, the Gaussian energy term is always constant and, thus, does not contribute to the energy balance. Hence, it can be neglected without any restriction. Another bending energy term related to the area difference of the inner and outer lipid monolayers of the RBC membrane may be included [31, 35, 40]. This non-local contribution is neglected in the present thesis for the sake of simplicity. Indeed, it is commonly disregarded by other scientists as well (e.g., [44, 76, 84, 86, 89]).

In order to find an *efficient* discretized bending energy on a triangular mesh (section 8.3), it is beneficial to start with a membrane without spontaneous curvature, $H^{(0)} = 0$, first. In this case, the energy E_B in eq. (7.8) is commonly discretized in the form [201, 202]

$$E_\mathrm{B} = \frac{\tilde{\kappa}_\mathrm{B}}{2}\sum_{\langle i,j\rangle}(\boldsymbol{n}_i - \boldsymbol{n}_j)^2 = \tilde{\kappa}_\mathrm{B}\sum_{\langle i,j\rangle}(1 - \boldsymbol{n}_i\cdot\boldsymbol{n}_j) = \tilde{\kappa}_\mathrm{B}\sum_{\langle i,j\rangle}(1 - \cos\theta_{ij}) \tag{7.9}$$

where the sum runs over all pairs $\langle i,j\rangle$ of neighboring faces[1] of the tessellated membrane. The unit normal vector of face i is $\boldsymbol{n}_i$, and the angle between neighboring normal vectors $\boldsymbol{n}_i$ and $\boldsymbol{n}_j$

[1] Two faces are neighbors if they share an edge, i.e., two nodes.

is θ_{ij}. The constant $\tilde{\kappa}_B$ has to be chosen in such a way that eq. (7.8) and eq. (7.9) are consistent for small angles. According to Gompper and Kroll [201], this is the case for $\tilde{\kappa}_B = \sqrt{3}\kappa_B$ for a membrane with the topology of a sphere. Assuming small angles, it is convenient to expand the cosine and approximate the bending energy by

$$E_B = \frac{\tilde{\kappa}_B}{2} \sum_{\langle i,j \rangle} \theta_{ij}^2. \tag{7.10}$$

At this point, a model assumption is made. In reality, the biconcave shape of the RBC results from the minimization of the bending energy subject to surface and volume constraints [31]. In the present model, however, this biconcave shape is used as *input*, i.e., it is assumed to be the correct equilibrium shape. Any deviation from this shape will then lead to an increase of the bending energy. This idea may be written in the form

$$E_B = \frac{\tilde{\kappa}_B}{2} \sum_{\langle i,j \rangle} \left(\theta_{ij} - \theta_{ij}^{(0)}\right)^2 = \tilde{\kappa}_B \sum_{\langle i,j>i \rangle} \left(\theta_{ij} - \theta_{ij}^{(0)}\right)^2. \tag{7.11}$$

For the undeformed membrane, each angle between neighboring faces 'remembers' its equilibrium value, and any deviation is penalized by an energy contribution to E_B. This approach also reflects the shape memory of RBCs [92]. The model used by Dupin et al. [87] bases on a similar idea although no sound bending energy is defined. The form of the energy in eq. (7.11) may also be motivated by assuming that the the energy is arbitrary but has a minimum at $\theta_{ij} = \theta_{ij}^{(0)}$. For small angle deviations $\delta\theta_{ij} = \theta_{ij} - \theta_{ij}^{(0)}$, the energy may be expanded about the minimum, taking only the leading term into account. Being heuristic, this approach is efficient in terms of computing time and circumvents the direct discretization of the curvature on the triangular mesh which is generally a delicate task.

The derivation of the bending forces,

$$\boldsymbol{F}_i^B = -\frac{\partial E_B(\{\boldsymbol{x}_i\})}{\partial \boldsymbol{x}_i}, \tag{7.12}$$

based on the energy in eq. (7.11) is presented in appx. C.2.

It should be noted that, in principle, simulations may be performed without bending resistance in order to simplify the model. Yet, a bending resistance has to be included whenever strong local curvatures appear. Else, the membranes can buckle or collapse [77, 84, 191, 203].

7.3. Membrane surface dilation energetics

In reality, the total surface area of a RBC is strongly conserved (section 3.1). Practically, one may introduce a surface energy [38, 204]

$$E_A = \frac{\kappa_A}{2} \frac{\left(A - A^{(0)}\right)^2}{A^{(0)}} \tag{7.13}$$

as penalty for surface deviations. The equilibrium surface area is $A^{(0)}$, and the current surface area is A. The magnitude of the surface energy is controlled by the surface modulus κ_A which is about $0.5\,\mathrm{N\,m^{-1}}$ for a RBC [31] and thus five orders of magnitude larger than the area dilation modulus of the cytoskeleton, $\kappa_\alpha = 5 \times 10^{-6}\,\mathrm{N\,m^{-1}}$ (section 7.1). The large surface modulus is caused by the incompressibility of the lipid bilayer which is basically a 2D fluid [31].

For simulations of RBCs, the surface area fluctuations should be as small as possible. Due to the explicit nature of the simulation algorithm, the numerical value for κ_A cannot be as large as the physical value, and a rigorous surface conservation cannot be realized. Yet, the observed surface deviations are usually smaller than 1%, even for strongly deformed RBCs and numerical values for κ_A which are about 1000 times smaller than in reality. A similar reduction of the numerical value for the surface modulus is also reported in [31].

The discretized form of the surface force acting on node i,

$$\boldsymbol{F}_i^{\mathrm{A}} = -\frac{\partial E_{\mathrm{A}}(\{\boldsymbol{x}_i\})}{\partial \boldsymbol{x}_i}, \tag{7.14}$$

is derived in appx. C.3.

7.4. Membrane volume energetics

In its undeformed shape, having the volume $V^{(0)}$, a RBC is in osmotic equilibrium with concentration c_0 of osmotically active molecules in its interior. The membrane is permeable for water but not for ions. Thus, a change in volume leads to a modified concentration and consequently to a modified osmotic pressure. This causes an increase of the osmotic free energy of the cell which can be approximated by [204]

$$E_{\mathrm{V}} = \frac{\kappa_{\mathrm{V}}}{2} \frac{\left(V - V^{(0)}\right)^2}{V^{(0)}} \tag{7.15}$$

and has the same form as the surface energy in eq. (7.13). Here, $\kappa_V = RTc_0$ is the volume (or osmotic) modulus where R is the universal gas constant and T is the temperature. The osmotic modulus for a RBC is $\kappa_V = 7.23 \cdot 10^5\,\mathrm{J\,m^{-3}}$ [31].

The energy scale related to the osmotic pressure is so large that any volume change of a RBC is always negligible, as long as the osmotic concentration of the ambient fluid is constant. Since, in the present thesis, temperature or ionic concentration fields are not considered, one may ask why a volume energy is introduced in the first place. This can be understood in the following way: Assuming simulations of impermeable membranes with no-slip condition at the surface and incompressibility of the interior and exterior fluids, the volume of the RBCs should be conserved at all times, even without enforcing this explicitly. However, for immersed boundary lattice Boltzmann simulations, both assumptions are only valid in the hydrodynamic limit. In a numerical simulation with finite spatial resolution, the no-slip condition cannot be exactly reproduced by the IBM—at least not for the originally proposed scheme which is used in this thesis [172, 205, 206, 207]. The reason is that, even if the fluid is incompressible and its velocity field divergence-free, the interpolated velocity field due to the IBM interpolations is not generally divergence-free. Furthermore, it is well-known that the LBM is a slightly compressible Navier-Stokes solver [208]. While the latter effect will only cause fluctuations of the cell volume about its initial magnitude, the former may lead to an accumulated volume drift, especially in long simulations with coarse lattice and mesh resolutions. In order to control the volume of the RBCs numerically, a volume energy of the form as in eq. (7.15) is used. It should be stressed that the volume energy here is employed to counteract a purely numerical shortcoming of the model, although its functional form can be motivated from physical considerations.

The discussion from section 7.3 can be applied to the volume energy as well: In simulations, unrealisticly small numerical values for κ_V have to be used for the sake of stability. Still, the volume deviations are sufficiently small, typically below 1%.

The derivation of the discretized volume force acting on node i,

$$\boldsymbol{F}_i^{\mathrm{V}} = -\frac{\partial E_{\mathrm{V}}(\{\boldsymbol{x}_i\})}{\partial \boldsymbol{x}_i}, \tag{7.16}$$

can be found in appx. C.4.

8. Advanced model discussions

In this chapter, advanced aspects of the computational model are presented which have not been addressed in one of the former chapters. A short overview of the sub-steps of the combined algorithm is given in section 8.1. The unit conversion between physical and lattice units is explained in section 8.2, followed by the description of the membrane mesh generation in section 8.3. The behavior of a single spherical, elastic capsule in shear flow is studied in detail in section 8.4 as a benchmark test. An efficient approach for initializing a dense suspension of deformable particles is discussed in section 8.5. Shortcomings and restrictions of the presented model are identified in section 8.6. The need for additional membrane-membrane interaction forces is motivated in section 8.7. Finally, wall slip at smooth walls and its circumvention is discussed in section 8.8.

8.1. Overview of the combined simulation algorithm

After the simulation has been initialized (sections 5.3.1 and 8.5), each time step of the combined algorithm as described in chapters 5, 6, and 7 consists of the sub-steps mentioned below. $\boldsymbol{x}_i$ denotes the position of the membrane node with index i, and $\boldsymbol{X}$ is the position of a fluid lattice node. The time step Δt is set to unity.

1. At the beginning of time step t, the membrane node positions $\boldsymbol{x}_i(t)$ and the fluid state $\boldsymbol{u}(\boldsymbol{X}, t)$, $\rho(\boldsymbol{X}, t)$ are known. From the configuration of the membranes, the forces $\boldsymbol{F}_i(t)$ acting on the membrane nodes in the Lagrangian frame are computed using the membrane model (chap. 7).
2. The membrane forces $\boldsymbol{F}_i(t)$ are spread to the Eulerian grid via immersed boundary method (IBM), cf. eq. (6.5), and the body force density $\boldsymbol{f}(\boldsymbol{X}, t)$ is obtained.
3. The body force $\boldsymbol{f}(\boldsymbol{X}, t)$ is used as input for the lattice Boltzmann method (LBM) which provides the new state of the fluid, $\boldsymbol{u}(\boldsymbol{X}, t+1)$, $\rho(\boldsymbol{X}, t+1)$ (chap. 5).
4. The new velocities of the membrane nodes, $\dot{\boldsymbol{u}}_i(t+1)$, are computed via IBM, cf. eq. (6.6).
5. The new positions of the membrane nodes, $\boldsymbol{x}_i(t+1)$, are found by evaluating eq. (6.12).
6. Information about the membrane and fluid states may be written to the disk for post-processing.
7. Go to sub-step 1 and proceed with time step $t+1$.

The sub-steps are also illustrated in fig. 8.1.

8.2. Conversion between physical and lattice units

A computer does not have the ability to compute dimensional quantities. Unit conversions are necessary in order to input physical data into a simulation and extract physical results again.

The first step is to realize that all mechanical quantities have a unit which can be decomposed into powers of length, time, and density. In the following, it is assumed that only mechanical quantities are required. Temperature and electric charges do not play a role in the present thesis.

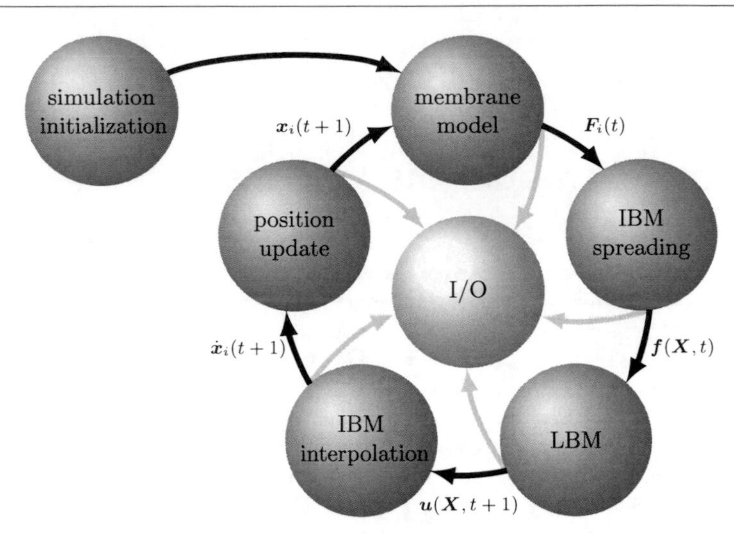

Fig. 8.1.: Scheme of the combined simulation algorithm. It proceeds along the black arrows. Data of each sub-step may be written to the disk (gray arrows) for post-processing. More details are given in section 8.1.

Each physical quantity Q may then be written in the form

$$Q = \tilde{Q} C_x^{q_x} C_t^{q_t} C_\rho^{q_\rho} \tag{8.1}$$

where $\tilde{Q}$ is a number (the value of Q in lattice units), and C_x, C_t, and C_ρ are the conversion factors for length, time, and density (with units m, s, and kg m^{-3}), respectively. The three numbers q_x, q_y, and q_ρ are the according and uniquely defined powers required to construct the correct unit for Q. Once the three conversion factors C_x, C_t, and C_ρ are known, units can be bidirectionally converted between the lattice and the physical system. It is straightforward to define derived conversion factors, e.g., for the pressure p. The unit of pressure is Pa = kg m^{-1} s^{-2} and thus $C_p = C_\rho C_x^2 C_s^{-2}$. The lattice and physical values of the pressure are then connected via $p = \tilde{p} C_p$. This example can be generalized to any other dimensional mechanical quantity. Numbers have the same values in lattice and physical units, i.e., the conversion factor is unity.

In lattice Boltzmann simulations, C_x is usually the physical length associated with the distance between two neighboring lattice nodes (the lattice constant Δx), $C_x = \Delta x$ and $\Delta \tilde{x} = 1$. This will always be the case in the present thesis. Since the lattice value for the density is commonly set to unity, $\tilde{\rho} = 1$, C_ρ automatically equals the density of the fluid, $C_\rho = \rho$. It remains the identification of the time conversion factor C_t.

If the Reynolds number is the only relevant dimensionless parameter in a lattice Boltzmann simulation, the time conversion factor is computed in the following way: Since the Reynolds number is dimensionless, its value must be the same in the lattice and physical systems. Else, the systems would not be equivalent. The kinematic viscosity of the fluid can then be written as

$$\nu = \tilde{\nu} \frac{\Delta x^2}{\Delta t} = \tilde{\nu} \frac{C_x^2}{C_t} \tag{8.2}$$

where $\Delta x = C_x$ has been used and $\Delta t \stackrel{!}{=} C_t$ (i.e., $\Delta \tilde{t} = 1$) is the unknown time step. A comparison with eq. (5.7) reveals that

$$\tilde{\nu} = \frac{\tau - \frac{1}{2}}{3} \tag{8.3}$$

is the viscosity in lattice units for a lattice with $c_s = \sqrt{1/3}\Delta x/\Delta t$. If the viscosity of the fluid, ν, the lattice constant, Δx, and the relaxation parameter, τ are known, the time step can be computed,

$$C_t = \Delta t = \frac{\tau - \frac{1}{2}}{3} \frac{\Delta x^2}{\nu}. \tag{8.4}$$

Eq. (8.4) can also be exploited to set up a simulation if another quantity than Δt is initially unknown.

As already mentioned in section 5.5, the LBM is not well suited to simulate low Reynolds number flows since the time step usually becomes small and a large number of time steps is required. This becomes also clear from eq. (8.4): A large viscosity ν leads to a small time step Δt. However, this raises the question whether the exact value of the Reynolds number (as long as it is small) is important after all. The time scale in highly viscous flows is arbitrary and does not depend on the Reynolds number [194, 209]. If the numerical Reynolds number was increased by a factor n, the time step would be increased by the same factor (if Δx and τ are not changed) and the number of required time steps and the related computing time would both be reduced by n. Of course, such an approach is only admissible if the physical results are not significantly compromised. Yet, a similar approach is constantly used in lattice Boltzmann simulations: The lattice Mach number is usually much larger than in reality. As long as the lattice Mach number is small, no *significant* effects are expected. It is generally not affordable to simulate fluids with the correct Mach number. Cates et al. [209] state that ' "fully" realistic simulations (in which lattice parameter values map directly onto those of the real world) are not the goal of mesoscale lattice Boltzmann.' In the present work, suspensions in the viscous regime are simulated, thus, the Reynolds number is small, and inertia effects are not relevant. For that reason, the time step may be increased as much as it is still compatible with the small Reynolds number assumption and as long as numerical stability is not endangered.

It is important to recognize that another dimensionless number is more important than the (small) Reynolds number when deformable particles in an external flow field are considered. The capillary number

$$\text{Ca} := \frac{\rho\nu\dot{\gamma}r}{\kappa_S} \tag{8.5}$$

quantifies the ratio of the viscous shear force of the fluid (density ρ, kinematic viscosity ν, and shear rate $\dot{\gamma}$) and the elastic shear force of the immersed particle (radius r, elastic shear modulus κ_S as introduced in section 7.1). The time scale of the physical problem is, therefore, defined by the capillary number, and the time conversion factor can be computed from

$$C_t = \frac{1}{C_{\dot{\gamma}}} = \frac{\tilde{\dot{\gamma}}}{\dot{\gamma}} = \frac{\text{Ca}}{\dot{\gamma}} \frac{\tilde{\kappa}_S}{\tilde{\rho}\tilde{\nu}\tilde{r}} \tag{8.6}$$

when the remaining parameters have already been chosen.

It is easy to see that the capillary number and the Reynolds number can be controlled independently. For example, a proportional increase of the lattice shear rate $\tilde{\dot{\gamma}}$ and the lattice shear elasticity $\tilde{\kappa}_S$ leaves the capillary number invariant, but the Reynolds number is increased since $\text{Re} \propto \dot{\gamma}$ but $\text{Re} \neq \text{Re}(\kappa_S)$. This freedom will be used to increase the time step Δt for a fixed and well-defined capillary number in chap. 10.

For the sake of simplicity, the tilde for indicating the dimensionless value of a quantity will be dropped in the remainder of this thesis, except explicitly noted otherwise. Unit conversions in LB simulations are also discussed by Ding and Aidun [154] and Feng et al. [166].

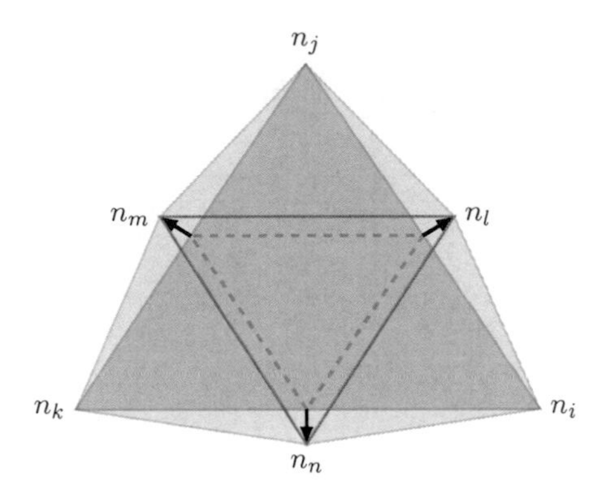

Fig. 8.2.: Membrane mesh generation by subdivision. Each face of the original icosahedron which is defined by three nodes n_i, n_j, and n_k (dark-gray) is subdivided into N^2 elements of equal size ($N = 2$ in this example). New nodes (here: n_l, n_m, n_n) are created and connected in such a way that N^2 faces of equal area are produced (dashed lines). Finally, the new nodes are radially shifted (out of the plane, black arrows) until they are located on the circumsphere of the icosahedron. The N^2 final faces are shown in light gray.

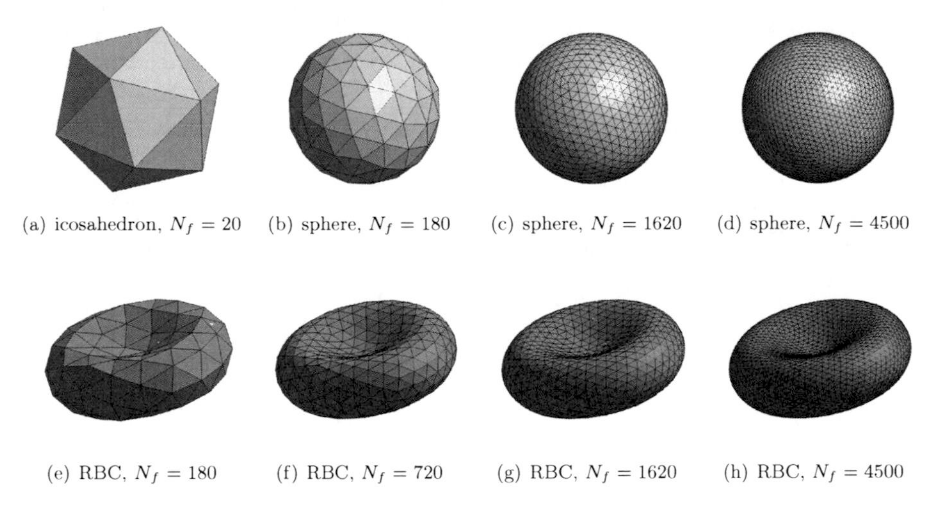

(a) icosahedron, $N_f = 20$ (b) sphere, $N_f = 180$ (c) sphere, $N_f = 1620$ (d) sphere, $N_f = 4500$

(e) RBC, $N_f = 180$ (f) RBC, $N_f = 720$ (g) RBC, $N_f = 1620$ (h) RBC, $N_f = 4500$

Fig. 8.3.: Examples of sphere and red blood cell meshes. The original icosahedron mesh with 20 faces ($N_f = 20$) is shown in (a). (b)–(d) Spherical meshes for various values of N_f. (e)–(h) Red blood cell meshes for various values of N_f.

8.3. Membrane mesh generation

There are basically two different ways to treat the Lagrangian mesh. First, one may introduce a structured, curvilinear grid with an intrinsic coordinate system (e.g., polar coordinates). Second, an unstructured grid (irregular decomposition of the surface into small patches whose connectivity must be specified explicitly) may be employed [210]. While the first approach usually leads to coordinate singularities at the poles, gradients have to be approximated in the second method. Both approaches are commonly used to describe deformable particles immersed in fluids. For example, Diaz et al. [211] and Lac et al. [212] have used structured meshes whereas Kraus et al. [76], Navot [192], and Ramanujan and Pozrikidis [191] have employed unstructured grids. In the present model, an unstructured mesh with triangular face elements is used as shown in fig. 8.3.

Tab. 8.1.: Properties of the sphere and red blood cell (RBC) meshes. The meshes for a sphere and a RBC for $N_f = 1620$ as shown in fig. 8.3(c) and fig. 8.3(g) are compared. A is the face area, l is the edge length (distance between neighboring nodes), φ is the interior angle in a face. The bar indicates the average of a quantity taken over the entire mesh, and σ denotes its standard deviation. The listed values are obtained for a spherical mesh with radius 7.5 and a RBC mesh with large radius 9 (both leading to $\bar{l} \approx 1$). All lengths and areas are given in lattice units.

quantity	sphere ($r = 7.5$)	RBC ($r = 9$)
range of face areas	$[0.40 \ldots 0.46]$	$[0.37 \ldots 0.50]$
average face area $\bar{A}$	0.43	0.44
average area deviation $\sigma_A/\bar{A}$	3.0%	7.5%
range of edge lengths	$[0.93 \ldots 1.08]$	$[0.80 \ldots 1.27]$
average edge length $\bar{l}$	1.01	1.01
average length deviation $\sigma_l/\bar{l}$	4.8%	9.2%
range of edge angles	$[53° \ldots 72°]$	$[41° \ldots 84°]$
average angle deviation $\sigma_\varphi/\bar{\varphi}$	8.0%	13.7%

Creating an unstructured mesh for a given surface (e.g., for a RBC or a sphere) is a non-trivial task. It is desirable to design meshes which are as homogeneous and isotropic as possible in order to minimize potential discretization artifacts. Feng and Michaelides [213] use a minimum potential approach, i.e., membrane nodes can move freely on the surface and interact via a repulsive pair potential. After some time, a node configuration is reached which corresponds to a local energy minimum. This configuration is then saved and used for later simulations. It is also possible to create triangular meshes directly with software tools like CGAL [214] or Gmsh [215]. Although simply obtained, these meshes seem to lack the desired homogeneity and isotropy (Krüger et al. [187]).

There is an approach to create high quality meshes for spherical surfaces, similar to that presented in [191]. One starts from a highly symmetric Platonic solid with N_f triangular faces. Here, an icosahedron with $N_f = 20$ is taken, cf. fig. 8.3(a). Each flat triangular surface element is then subdivided into N^2 equisized triangular sub-elements as indicated in fig. 8.2. The new nodes are radially shifted to the circumsphere of the icosahedron. This approach guarantees a surpassing homogeneity and isotropy of the mesh which is of large importance for the model, as discussed in section 8.6. Any closed triangular mesh with N_f faces has N_n nodes where $N_f = 2N_n - 4$. For a mesh created from an icosahedron as stated above, each node is member of five or six faces: The original 20 icosahedron nodes are member of five faces, all remaining nodes of six faces.

For an existing spherical mesh, it is straightforward to obtain the corresponding mesh for a RBC. The average shape of a RBC under physiological conditions can be parameterized by [216]

$$z(\varrho) = \pm\sqrt{1 - \left(\frac{\varrho}{r}\right)^2} \left(C_0 + C_2 \left(\frac{\varrho}{r}\right)^2 + C_4 \left(\frac{\varrho}{r}\right)^4\right). \tag{8.7}$$

The rotational symmetry axis of the RBC is along the z-axis ($\varrho = \sqrt{x^2 + y^2}$), and the parameters read $r = 3.91\,\mu\text{m}$ (large radius of a RBC), $C_0 = 0.81\,\mu\text{m}$, $C_2 = 7.83\,\mu\text{m}$, and $C_4 = -4.39\,\mu\text{m}$. Each point $\boldsymbol{x}'$ of the spherical mesh is then shifted to the point $\boldsymbol{x}$ of the RBC mesh according to $\boldsymbol{x}' = (x', y', z') \to \boldsymbol{x} = (x = x', y = y', z = z(\varrho'))$. Some exemplary meshes are shown in fig. 8.3.

Meshes are created in advance and saved as input data files for the simulations. Since the topology of the Lagrangian meshes and the connectivity of the membrane nodes never change, there is no need for any remeshing during a simulation. The high quality of the produced meshes can be inferred from fig. 8.3 and tab. 8.1: All face areas are of comparable size, and the edge length distribution has only a small width. Neither extremely small nor large face interior angles appear. This is important for the stability and a reduction of numerical artifacts as explained in section 8.6.

8.4. Benchmark test: single capsule in shear flow

This section bases on the investigations published in Krüger et al. [187]. In order to benchmark the combined algorithm for the fluid, the membrane, and their mutual coupling, a series of test simulations has been performed. The main intention of this benchmark is to understand (i) the numerical effect of the IBM interpolation stencil and (ii) the significance of the ratio $\bar{l}/\Delta x$ of capsule mesh and fluid lattice resolutions. Here, $\bar{l}$ is the average distance between neighboring membrane nodes (section 8.3), and Δx is the lattice constant (section 5.2). It is of primary interest to reveal how large the uncertainties for small or intermediate spatial resolutions are. Since a large number of particles shall be simulated eventually, the affordable resolution is restricted.

A single capsule with spherical rest shape and radius r is placed in the middle of a simple shear flow with external shear rate $\dot{\gamma}$ induced by moving walls at $z = \pm L_z/2$ with velocities $\pm u_{\mathrm{w}}$ along the x-axis (thus, $\dot{\gamma} = 2u_{\mathrm{w}}/L_z$). The system is cubic with size L_z^3. Due to the shear flow, the particle deforms and rotates about its stationary origin. After an initial transient, a steady configuration develops which is illustrated in fig. 8.4. The particle shape is a stationary, inclined ellipsoid whereas the membrane itself is 'tank-treading' with constant angular velocity ω about this shape. From this, two relevant observables can be extracted: the inclination angle θ and the deformation parameter D of the capsule. The inclination angle θ is defined as the angle between the flow axis and the largest semiaxis of the ellipsoid. The deformation parameter is

$$D := \frac{a - c}{a + c} \tag{8.8}$$

where a and c are the largest and smallest semiaxes of the deformed capsule, respectively. For a sphere, $D = 0$ holds. The angular velocity ω may also be computed (Krüger et al. [187]), but it is not in the focus of this benchmark.

Finding analytic solutions for problems involving deformable particles in external flows is not trivial. This makes it difficult to benchmark the IBM applied to deformable particles. However, for the above-mentioned problem, there exists an analytical solution [217] if (i) the constitutive elastic law of the membrane is known, (ii) the Reynolds number $\mathrm{Re} = \dot{\gamma}r^2/\nu$ of the flow can be neglected, (iii) the deformation of the capsule is small, $D \ll 1$, and (iv) the external flow is unbounded shear flow, i.e., the walls are far away, $L_z \gg r$. In the present case, the internal and external Newtonian fluids have the same density and viscosity. The capsule is only subject to the in-plane shear forces as described in section 7.1 because the theory does not include bending forces or non-local surface and volume forces.

When Skalak's constitutive law with $\kappa_{\mathrm{S}} = \kappa_\alpha$ is used (section 7.1), theory predicts $\theta_{\mathrm{opp}}/\pi := (\pi/4 - \theta)/\pi = 15\mathrm{Ca}/8$ for the inclination angle and $D = 25\mathrm{Ca}/4$ for the deformation parameter in steady state [187, 217] where

$$\mathrm{Ca} = \frac{\rho\nu\dot{\gamma}r}{\kappa_{\mathrm{S}}} \tag{8.9}$$

is the capillary number of the membrane, cf. eq. (8.5). For convenience, the angle $\theta_{\mathrm{opp}} \propto \mathrm{Ca}$ has been defined. It is the angle between the largest semiaxis of the capsule and the main diagonal in the shear plane. Obviously, in the limit of vanishing deformation, the inclination angle θ is $\pi/4$, i.e., 45°.

Preparations

Before the actual benchmark simulations can be performed, reasonable simulation parameter values have to be found. In particular, it has to be investigated (i) up to which Reynolds number the Stokes approximation is satisfied, (ii) which ratio of box size and particle radius H/r is

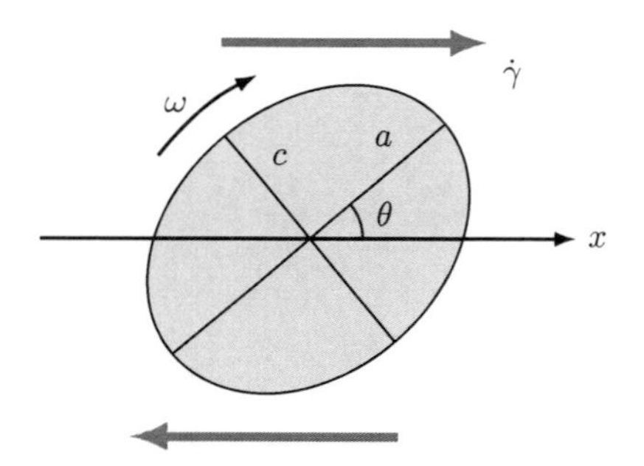

Fig. 8.4.: Tank-treading capsule. The capsule cross-section is shown in the xz-plane. It is deformed with major and minor semiaxes a and c. The inclination angle θ is taken between the major semiaxis and the x-axis (velocity direction of the external flow, dark gray arrows). The membrane rotates about its spatially fixed shape with angular velocity ω ('tank-treading').

required for *unbounded* shear flow, (iii) which deformation parameter D still is 'small', and (iv) which value of the LBM relaxation parameter τ is reasonable. The quadratic equilibrium distribution, eq. (5.6), is taken for all simulations in this section albeit the linearized equilibrium may have been used since the Reynolds number is small. The numerical fluid density ρ is set to unity throughout this section.

All of the preliminary tests have been thoroughly performed and discussed in Krüger et al. [187]. Only the results shall be given in the following. For these simulations, the 4-point interpolation stencil ϕ_4, eq. (6.8), has been used. The capsule radius is $r = 5\Delta x$, and the spherical, icosahedron-based mesh, cf. section 8.3, with $N_f = 1280$ faces ($\bar{l}/\Delta x = 0.76$) has been employed.

Although the transient behaves differently, the steady state value of the deformation parameter does not depend on the Reynolds number up to values of about 0.1 (larger values have not been tested). A relative box size of $L_z/r = 10$ was found to be sufficient to simulate unbounded shear flow. For Ca $= 0.01$ (corresponding to $D \approx 0.06$), the deformation is still 'small', and the linearized solution is sufficiently accurate for the problem description. Finally, it was found that the LBM relaxation parameter τ should not be larger than unity because higher values detrimentally affect the accuracy. Concluding, for the upcoming benchmark tests, the parameters have been fixed in the following way: Re $= 0.02$, $L_z/r = 10$, Ca $= 0.01$, $\tau = 1$. Having set these values, the only free parameters are the particle radius r in lattice units, the ratio $\bar{l}/\Delta x$ of membrane and lattice resolutions (and thus the capsule mesh size N_f), and the choice of the IBM interpolation stencil (ϕ_2, ϕ_3, or ϕ_4).

Convergence for fixed mesh resolution N_f

In this simulation series, the influence of the hydrodynamic resolution is tested alone, i.e., the capsule mesh resolution N_f is kept constant. This way, it is possible to study the effect of a non-constant ratio $\bar{l}/\Delta x$ by varying the capsule radius r. The employed mesh resolutions are $N_f = 320$ and 1280.

For the mesh with 320 faces, the capsule radius has been set to $r/\Delta x = 3$, 4, 5, and 6, corresponding to $\bar{l}/\Delta x = 0.90$, 1.20, 1.50, and 1.80, respectively. The results are shown in fig. 8.5. Although the mesh resolution N_f is small, it can be seen that the physics of the system is roughly captured. The accuracy of the solutions increases when the radius r becomes larger. Furthermore, it can be inferred that the values obtained with the interpolation stencil ϕ_2 are closest to the expected values. The larger the range of the IBM interpolation, the larger the deformation

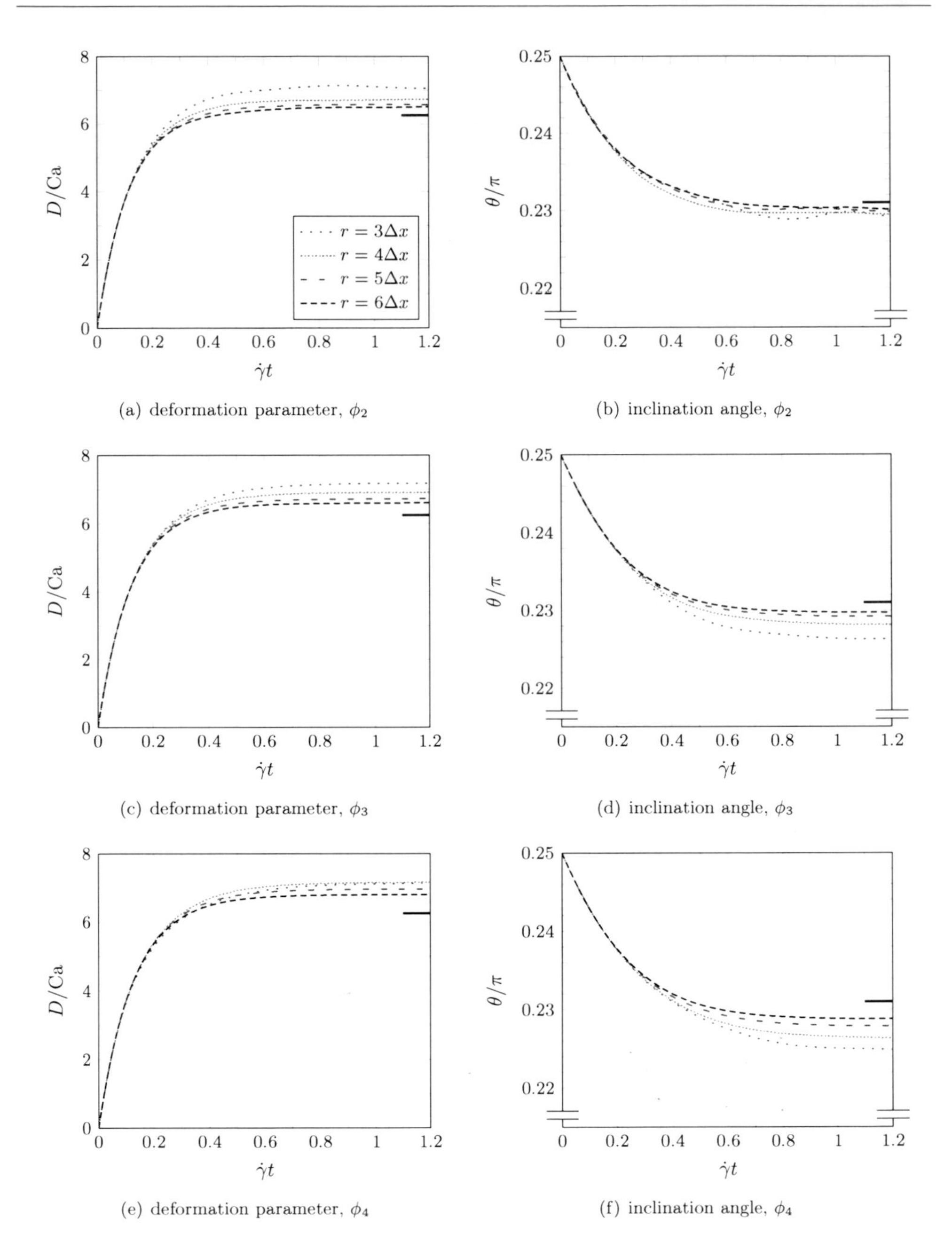

(a) deformation parameter, ϕ_2

(b) inclination angle, ϕ_2

(c) deformation parameter, ϕ_3

(d) inclination angle, ϕ_3

(e) deformation parameter, ϕ_4

(f) inclination angle, ϕ_4

Fig. 8.5.: Deformation parameter D and inclination angle θ for a tank-treading capsule. D/Ca and θ/π are shown for varying ratios $\bar{l}/\Delta x$ for a mesh with $N_f = 320$ faces and interpolation stencils ϕ_2 in (a) and (b), ϕ_3 in (c) and (d), and ϕ_4 in (e) and (f). $r/\Delta x = 3$, 4, 5, and 6 correspond to $\bar{l}/\Delta x = 0.90$, 1.20, 1.50, and 1.80, respectively. The expected values are shown as short black lines (6.25 for D/Ca and 0.231 for θ/π). The legend in (a) is valid for all subfigures.

parameter D and the larger (smaller) the angle θ_{opp} (θ). As detailedly discussed in Krüger et al. [187], this can be interpreted in the following way: The finite-range interpolations due to the IBM lead to an apparent growth of the capsule radius r, which increases the apparent capillary number Ca, cf. eq. (8.5). Thus, the deformation is slightly larger than expected. Consequently, the deformation excess is smallest for the narrow interpolation stencil ϕ_2. When the radius r is increased, the relative width of the interpolation decreases and the numerical results approach the analytical solution. The mesh with 1280 faces has been tested for radii $r/\Delta x = 3$, 5, 7, and 9, corresponding to $\bar{l}/\Delta x = 0.45$, 0.75, 1.06, and 1.36, respectively. The plots are not shown here separately[1] because the results are qualitatively similar to those obtained for $N_f = 320$: The results become more accurate when (i) the IBM interpolation stencil has a smaller support and (ii) when the radius r increases.

It has been observed that the 2-point interpolation stencil ϕ_2 produces significantly wrong solutions when $\bar{l}/\Delta x > 2$ (data not shown). At this point, the spacing between neighboring mesh nodes is so large that fluid can penetrate the capsule membrane without experiencing the no-slip condition. For ϕ_3 and ϕ_4, a similar behavior at $\bar{l}/\Delta x = 2$ has not been observed. The probable explanation is the larger range of the interpolations, still keeping the fluid from passing through the membrane. The exact value of the mesh ratio $\bar{l}/\Delta x$ seems to play only a minor role as long as it is not too small (< 0.5) or too large (> 1.5). This indicates that the resolution of the membrane—at least for small deformations—does not require an extremely fine mesh, and the mesh ratio $\bar{l}/\Delta x$ can be safely chosen somewhere between 0.5 and 1.5 without significantly influencing the physical results. This is an important result since it allows a certain flexibility in setting up the simulations, especially in view of efficiency. However, it has been seen that the ratio of interpolation width (caused by the IBM interpolation stencil) and the radius r of the capsule should be as small as possible. This point will also be discussed in the following.

Convergence for fixed mesh ratio $\bar{l}/\Delta x$

In this second series, both the mesh and the hydrodynamic resolutions are increased by the same rate, i.e., the mesh ratio $\bar{l}/\Delta x$ is fixed. The mesh and hydrodynamic resolutions are $N_f = 1280$ and $L_z = 35$, $N_f = 5120$ and $L_z = 70$, and $N_f = 20480$ and $L_z = 140$, respectively. The mesh ratio is $\bar{l}/\Delta x = 0.53$ in all cases. It has to be noted that for $N_f = 320$, a mesh ratio of $\bar{l}/\Delta x = 0.53$ leads to quite unacceptable results since the capsule radius r becomes too small compared to the numerical width of the interpolation stencils. The results are shown in fig. 8.6. It is obvious that the steady state values of D/Ca and θ/π converge to their analytic values (6.25 and 0.231, respectively) when the resolution is refined. In order to quantify the results, the errors at $\dot{\gamma}t = 1.2$ (which is already in steady state) are listed in tab. 8.2 and shown in fig. 8.7. The convergence is close to second order. The only exception is the convergence of the inclination angle θ with the interpolation stencil ϕ_2. This deviation is caused by mesh degradation: Since the applied forces are merely in-plane forces, the capsule mesh starts to form ripples after some time. These ripples are caused by numerical artifacts which are most significant for ϕ_2 since this interpolation stencil is not as smooth as ϕ_3 or ϕ_4 (fig. 6.3). In-plane forces cannot counteract the ripples, and a bending force would be required to avoid them (section 7.2). However, as mentioned before, a bending force cannot be included since the analytical solution does not account for it. Fortunately, in other simulations within this work, the bending force is always included, and ripple formation has never been observed in these cases.

Conclusions

It has been found that the finite range of the IBM interpolations apparently increases the membrane radius, which causes numerical inaccuracies. Compared to these, fluid lattice and

[1] They can be found in Krüger et al. [187] instead.

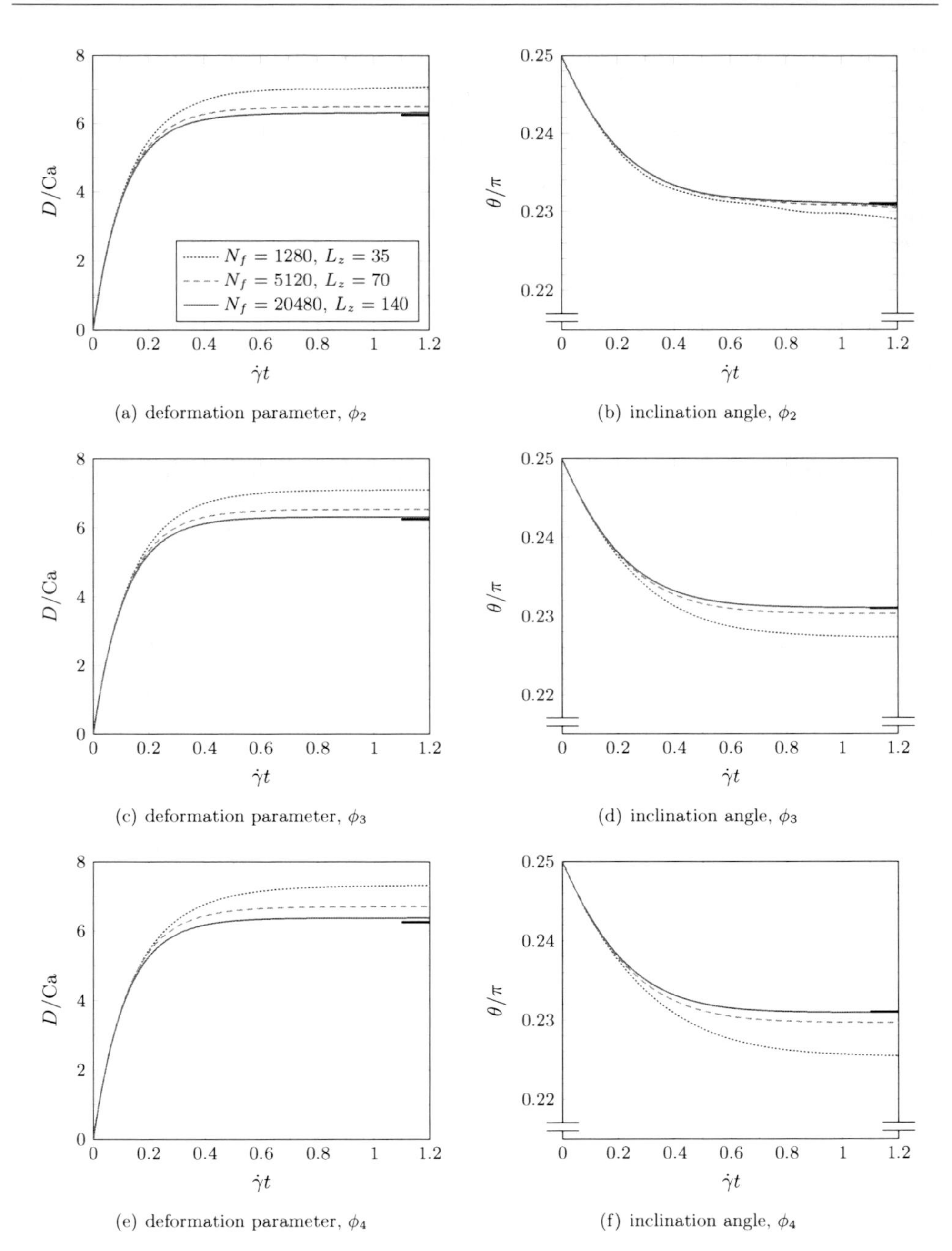

Fig. 8.6.: Deformation parameter D and inclination angle θ for a tank-treading capsule. D/Ca and θ/π are shown for varying resolutions L_z and N_f for $\bar{l}/\Delta x = 0.53$ and interpolation stencils ϕ_2 in (a) and (b), ϕ_3 in (c) and (d), and ϕ_4 in (e) and (f). The mesh and fluid resolutions are $N_f = 1280$ and $L_z = 35$ (dotted lines), $N_f = 5120$ and $L_z = 70$ (dashed lines), and $N_f = 20480$ and $L_z = 140$ (solid lines), respectively. The expected values are shown as short black lines (6.25 for D/Ca and 0.231 for θ/π). The legend in (a) is valid for all subfigures.

Tab. 8.2.: Convergence of the capsule's deformation parameter D and the angle θ_{opp} for the interpolation stencils ϕ_2, ϕ_3, and ϕ_4. The relative deviations $\delta D/D^{\text{a}} := (D^{\text{s}} - D^{\text{a}})/D^{\text{a}}$ and $\delta\theta_{\text{opp}}/\theta^{\text{a}}_{\text{opp}} := (\theta^{\text{s}}_{\text{opp}} - \theta^{\text{a}}_{\text{opp}})/\theta^{\text{a}}_{\text{opp}}$ are shown at time $\dot{\gamma}t = 1.2$ (subscripts 's' and 'a' denote simulation and analytical, respectively). The convergence order α is taken from a fit to the functions $\delta D/D^{\text{a}}, \delta\theta_{\text{opp}}/\theta^{\text{a}}_{\text{opp}} \propto L_z^{-\alpha}$. For $\delta\theta_{\text{opp}}$ and ϕ_2, a meaningful convergence order could not be obtained due to mesh degradation. A graphic representation of this table is shown in fig. 8.7.

resolution		ϕ_2		ϕ_3		ϕ_4	
L_z	N_f	$\frac{\delta D}{D^{\text{a}}}$	$\frac{\delta\theta_{\text{opp}}}{\theta^{\text{a}}_{\text{opp}}}$	$\frac{\delta D}{D^{\text{a}}}$	$\frac{\delta\theta_{\text{opp}}}{\theta^{\text{a}}_{\text{opp}}}$	$\frac{\delta D}{D^{\text{a}}}$	$\frac{\delta\theta_{\text{opp}}}{\theta^{\text{a}}_{\text{opp}}}$
35	1280	13.2%	12.0%	13.5%	20.9%	17.0%	30.8%
70	5120	4.1%	4.5%	4.5%	4.9%	7.3%	8.5%
140	20480	1.2%	3.0%	1.0%	0.9%	2.0%	1.7%
convergence order α		1.7	N/A	1.9	2.2	1.5	2.1

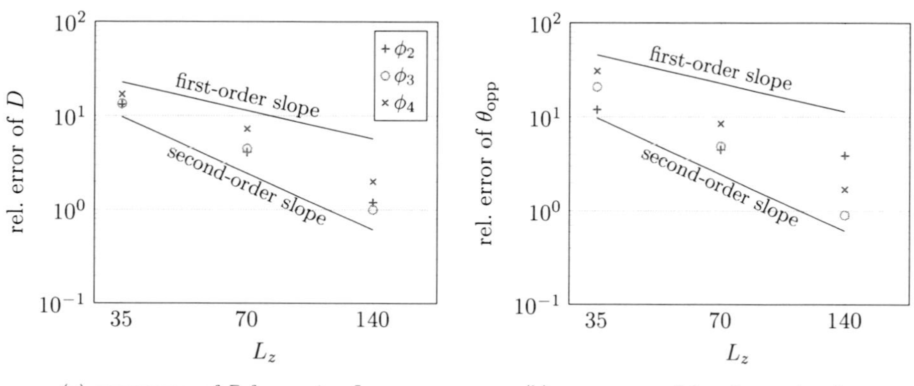

(a) convergence of D for varying L_z (b) convergence of θ_{opp} for varying L_z

Fig. 8.7.: Convergence of the deformation parameter D and the angle θ_{opp} for the interpolation stencils ϕ_2, ϕ_3, and ϕ_4. In (a), the relative error of the deformation parameter D is shown for increasing mesh and fluid resolutions with fixed $\bar{l}/\Delta x = 0.53$. The analog results for the relative error of the angle θ_{opp} are shown in (b). The data is taken from tab. 8.2. The legend in (a) is valid for (b) as well.

membrane meshing artifacts can be neglected, at least for small deformations. A large hydrodynamic resolution, i.e., a large value of $r/\Delta x$ is the only way to significantly reduce the numerical error. The computing time sets an upper bound for reasonable values of $r/\Delta x$. It is convenient to introduce a hydrodynamic radius r^* based on the true deformation D which is larger than the expected deformation. For the two-point interpolation stencil, the hydrodynamic radius is about $r^* \approx r + 0.3\Delta x$, which can be obtained from tab. 8.2. This observation will be important in chap. 10 where the simulation results for blood viscosity are compared to experimental data.

For a fixed mesh ratio $\bar{l}/\Delta x$, convergence to the analytical solution can be observed when r and N_f are simultaneously increased. The convergence order is close to two. This is convincing evidence that the presented numerical tool produces reliable results in the small deformation limit for which analytical solutions exist.

It has been seen that the resolution ratio $\bar{l}/\Delta x$ can be selected somewhere between 0.5 and 1.5 without compromising the physical results. This offers a flexible parameter choice with respect to the lattice and mesh resolutions. If $\bar{l}/\Delta x$ becomes smaller than 0.5, the IBM interpolations do not produce reliable results as will be explained in section 8.7. The present benchmarks are only valid for small deformations. For simulations with large deformations, however, the membrane mesh resolution must be sufficiently high to handle regions with large local curvature.

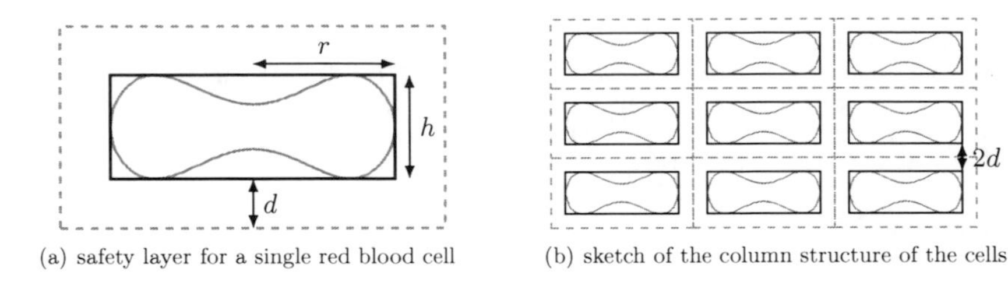

(a) safety layer for a single red blood cell (b) sketch of the column structure of the cells

Fig. 8.8.: Column initialization of a red blood cell (RBC) suspension simulation. (a) Each RBC is bounded by a box with height h and length/width $2r$ (solid rectangle). The bounding box is surrounded by a safety layer of thickness d (dashed line). (b) The mutual distance between any two neighboring bounding boxes is $2d$.

8.5. Initialization of dense suspensions: distribution of the particles in the simulation box

It is a formidable task to initialize a simulation of a dense suspension of particles. Here, 'dense' means volume fractions above 30–40%. Due to the generally non-spherical shape of the particles, in particular for RBCs, it is hard to fill the volume efficiently. It is also of significant importance to initialize the particles randomly to promote mixing and to reduce the transient time at the beginning of the simulations. In the following, two approaches for the RBC positioning are presented: column initialization and random initialization. The discussions can be easily generalized to other particle shapes as well.

Initializing particle positions in columns

The simplest way to arrange the RBCs is to put them in simple rectangular columns with a given gap between them. One can show that the obtainable hematocrit is

$$\mathrm{Ht} = \frac{4r^2h}{(2r+2d)^2(h+2d)} \times 58\% \tag{8.10}$$

where r is the large radius of the RBC, $h \approx \frac{2}{3}r$ is its height, and d is a safety layer around each cell (fig. 8.8). The numerical value of 58% in eq. (8.10) is the ratio of the volumes of a RBC and the tight bounding box surrounding it (for a sphere, this value is 52%). The shortest distance between two adjacent cells is $2d$. This safety layer is important as will be discussed in section 8.6. The maximum hematocrit which can be reached this way is 58% when $d \to 0$. However, d should not fall much below 0.5 lattice units. For a typical radius of $r = 8$ and $d = 0.5$, the hematocrit would be 43% and close to the physiological value of about 45%. In principle, the hematocrit could be increased by going to higher resolutions (increasing r) while keeping d constant. This approach, however, is computationally expensive.

The above method is not a good choice for dense suspensions since it requires a long transient simulation time to get rid of the unphysical column information. In order to break the periodicity of the columns, each cell may be shifted and rotated by a random value. Due to the necessary presence of the safety distance, these random shifts and rotations must be small (a few 0.1 lattice nodes and a few degrees, respectively). Otherwise the cells would come too close or may even overlap. Additionally, the simulation box must have dimensions which are multiples of $(2r + 2d)$ and $(h + 2d)$, respectively, if the volume should be efficiently filled. This is not possible in general. Any deviation from the rectangular geometry would lead to a reduction of the available hematocrit. Another possible way to break the periodicity is to remove single cells

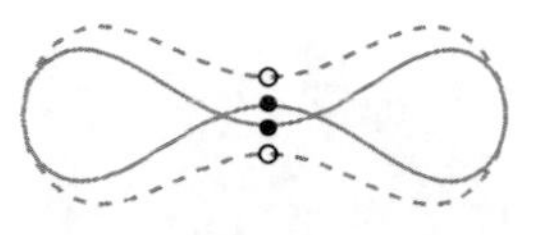

Fig. 8.9.: Self-intersection of a red blood cell (RBC) during initialization. If no internal fluid is used during the growth process, the top and bottom dimples (small circles) are not aware of each other, and the RBC may self-intersect (solid line, black circles). The undeformed shape is shown as comparison (dashed line, white circles). Introducing a repulsion force between the dimples solves this numerical problem.

from the lattice, i.e., to create gaps which accelerate randomization [87]. However, this reduces the hematocrit at the same time and is not recommendable for dense suspensions.

Due to the above-mentioned disadvantages, another approach must be considered for dense systems with Ht > 30%.

Initializing particle positions randomly

A promising way to position the RBCs randomly is pointed out by MacMeccan [194] and Clausen et al. [218]. The approach followed here is slightly different. It consists of two steps:

1. Position the cells with random locations and orientations. Use particles with 50% of their actual radius (initial hematocrit is one eighth of the desired value).
2. Grow the cells to their full size taking into account interaction forces to avoid jamming and overlap.

Randomly positioning particles with high volume fraction without overlap can be extremely time consuming since at some point most configurations do not allow of the addition of further particles. For this reason, the particles are first decreased in size and then randomly positioned throughout the simulation box. During positioning, it is checked whether any particle overlaps with a solid wall or any other particle which has already been created. Here, the cells are treated as spheres, and only the center-to-center distance is used for checking. If the new particle overlaps with a wall or any other particle already positioned, a new position vector is generated, and it is tested again. The positioning step is completed almost instantaneously since the effective volume fraction at this stage is typically only 5–8% and overlaps are rare.

The growing step is more challenging, and there are different approaches available. For the present thesis, the volume of the cells is increased from its reduced value V_0 (which is typically $V/8$) by a constant value V_+ at each time step. At each growth step, the membrane interaction forces as described in section 8.7 are evaluated. This way, overlap of the particles during growth is circumvented. The basic idea is to reach a high volume fraction by letting the particles deform during growth, i.e., they are allowed to change their shape in order to fill the available volume more efficiently. This approach differs from that presented in [194, 218] where the cells are rigid during growth. In order to maintain smooth membrane shapes, also the shear and bending forces (cf. sections 7.1 and 7.2) are computed at each growth step. It has turned out that 5000–10000 growth steps are sufficient for suspensions with volume fractions up to 70% and average particle radii of about 8 lattice units.

Hydrodynamic interactions are not required in this (unphysical) stage, and LBM and IBM are not used. Instead, a simple molecular dynamics model is employed for the integration of the equations of motion. The velocity and position of each node i at growth step $n + 1$ is computed

(a) after random positioning (b) after half of the process (c) after completed process

Fig. 8.10.: Random initialization of a dense red blood cell (RBC) suspension simulation. 520 RBCs with Ht = 46% are grown from their initial to their final size (initial radius is 50% of final radius). The cells are shown (a) directly after random positioning (Ht = 6%), (b) after half of the growth process (Ht = 23%), and (c) directly after the completed growth process.

from its velocity and position at step n and the total force it experiences at step $n+1$,

$$\begin{aligned} \dot{\boldsymbol{x}}_i(n+1) &= \dot{\boldsymbol{x}}_i(n) + \frac{1}{m}\boldsymbol{F}_i(n+1), \\ \boldsymbol{x}_i(n+1) &= \boldsymbol{x}_i(n) + \dot{\boldsymbol{x}}_i(n+1). \end{aligned} \tag{8.11}$$

This simplified dynamics has proven to be sufficient for the growth process. A typical value for the mass is $m = 100$ (lattice units). This value can be chosen arbitrarily as long as the growth process is stable. It is shown in fig. 8.10 for a RBC suspension with 46% volume fraction.

In order to improve the quality of the growth process, an additional force is required. Areas of the same membrane which are not directly connected are not aware of each other, i.e., the two dimples of a single RBC are mutually invisible. The reason is the locality of the shear and bending forces. In the full simulations, the dimples of a cell cannot touch each other since the interior fluid prevents them from doing so. In the growth process, due to the absence of the fluid coupling, the dimples may self-intersect (fig. 8.9). For that reason, an additional force is used during growth which introduces a repulsion of the nodes in the top and bottom nodes near the dimples. The exact form of this force is not relevant as long as it maintains the desired safety distance between the dimples.

8.6. Limitations and restrictions of the numerical model

In this section, the restrictions and limitations of the numerical model are presented. The focus lies on the issues related to the present work rather than on an exhaustive overview. The section is divided into three parts: The problems related to the LBM are briefly presented, followed by a more elaborate discussion about the IBM and some statements about the membrane model. The intention is to identify the simulation parameters and physical applications for which the model produces reliable results.

Restrictions of the lattice Boltzmann method

For the simulation parameters chosen in the present thesis, there are basically no numerical problems related to the LBM. Due to the small Reynolds numbers, the relaxation parameter τ is always chosen sufficiently large ($\tau \approx 1$), and stability problems do not occur. The lattice Mach number is usually kept below 0.1. A noticeable disadvantage of the LBM is the coupling of the discretizations of position and velocity space. As discussed in sections 5.5 and 8.2, the time step is already set when the Reynolds number and the spatial resolution have been chosen for a given value of τ. For small Reynolds number flows, this usually leads to extremely small time steps and long integration times. Hence, a redefinition of the time step as discussed in section 8.2 may be necessary.

Restrictions of the immersed boundary method

The simplicity of the IBM (both the concept and implementation) does not come without a price, and the devil is in the details. There are a few issues which have to be considered when the IBM is used for the simulation of dense suspensions.

It is commonly observed that the IBM does not properly work when membrane nodes (either belonging to the same membrane or to distinct membranes) come too close (e.g., [89]). The reason is the indirect position update: Membrane nodes are advected by the ambient fluid velocity only, and interpolations of the velocities are required. If the distance between two membrane nodes is significantly smaller than the lattice constant Δx, velocity gradients do not survive the interpolation, and the nodes move with similar velocities. Consequently, it is not possible to provide a large velocity gradient for nearby membrane nodes, and these nodes cannot be separated in a realistic time, even if there is a large repulsion force between them. Membranes, thus, can stick together, in the worst case for all times. This reasoning is explained in more detail in section 8.7. It is obvious that the above limitation is more visible in dense suspensions than in dilute systems since the membrane node density is larger then. A high mesh quality significantly helps to minimize this problem: If the mesh is designed in such a way that the node-node distance distribution is narrow and centered at about Δx (cf. section 8.3), nodes within a given membrane usually never come too close. It remains to deal with the problem of membrane-membrane collisions, as discussed in more detail in section 8.7.

The above problem gives a somewhat natural answer to the question how to choose the ratio of Eulerian grid and Lagrangian mesh resolutions. In principle, the average node-node distance in a membrane, $\bar{l}$, can be set independently of the lattice constant Δx. It is expected that there is a given ratio $\bar{l}/\Delta x$ for which the numerical model works most reliably. On the one hand, if $\bar{l}/\Delta x$ is too small, the nodes stick together since the Eulerian grid cannot provide a sufficient resolution for the velocity field (see above and section 8.7). On the other hand, too large a value for $\bar{l}/\Delta x$ will lead to 'holes' in the membrane, cf. fig. 8.11(a). Fluid may then penetrate the membrane without being forced by nearby membrane nodes. The no-slip condition will then be violated and the numerical results become less accurate. Thus, the distance between neighboring membrane nodes should not be larger than the range of the force spreading defined by the width of the interpolation stencils in section 6.3. Based on these considerations and the results in Krüger et al. [187], the average node-node distance is set to $\bar{l} \approx \Delta x$ in all upcoming simulations if not otherwise stated.

A problem similar to the node-node behavior at small distances appears when nodes come too close to a solid wall. If a wall node is within interpolation range of a membrane node, on the one hand, force is spread to the wall and therefore 'lost' with respect to the fluid. On the other hand, it is not generally clear which velocity the wall should be assigned to in order to obtain the correct interpolated node velocity, cf. fig. 8.11(b). Bagchi [88] has circumvented the problem by assuring that the cells always have a minimum distance from the walls. Feng and Michaelides [79] introduce a repulsion force so that nodes never come too close to the wall. A similar path is also followed in the present thesis (section 8.7). It is noteworthy that analog problems also arise in other numerical methods where Eulerian and Lagrangian systems are combined (e.g., [85, 219]).

The time steps for the LBM and the membrane update are the same since they are directly

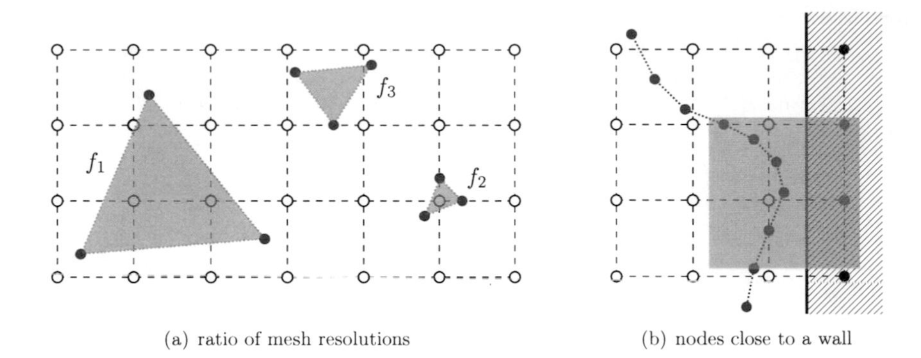

Fig. 8.11.: Problems related to the immersed boundary method. (a) Ratio of mesh resolutions: Three faces (red triangles) with different node-node distances $\bar{l}$ are shown (face f_1: $\bar{l} = 2.4\Delta x$, face f_2: $\bar{l} = 0.5\Delta x$, face f_3: $\bar{l} = 0.9\Delta x$). While the node-node distances for face f_1 are so large that lattice nodes (white circles) may not feel a membrane force at all, the nodes in face f_3 are so close that the interpolation becomes inaccurate. For face f_3, both detrimental effects are minimized. (b) Nodes close to a wall: If nodes (red circles) of a membrane (curved dotted line) are in direct wall proximity, force has to be spread to wall nodes (black circles within square region). Conversely, the velocity interpolation requires velocity information from the wall.

coupled via IBM, cf. section 8.1. Polymers immersed in a fluid and subject to thermal fluctuations may be simulated via LBM, molecular dynamics (MD), and viscous coupling [182]. In this case, a separation of time steps for the polymer (MD) and the fluid (LBM) is required. It is not unusual to have more than 50 MD time steps within one LBM time step. In the present case, this is not necessary since thermal fluctuations are not considered and the membrane motion is overdamped. Additionally, it is often stated that the explicit time integration of the IBM leads to numerical instabilities, and improvements have been proposed [220, 221]. This problem, however, is usually observed in systems of rigid particles. These particles may be modeled as deformable objects with high rigidity. Any deformation is penalized by a strong response force trying to maintain the rigid shape, which can lead to unphysical oscillations. On the contrary, in the present case, the particles are designed to be deformable, i.e., the membrane moduli are small (except for the surface and the volume conservation, see below), and large deformations lead to manageable force magnitudes.

Similar to nearly all Eulerian-Lagrangian coupling methods, the IBM formally is only first order accurate in space when sharp interfaces (e.g., membranes) are simulated [78]. Peng and Luo [222] observed that that the IBM has a second-order convergence behavior for LBM fluid flow around a rigid cylinder in 2D. Contrarily, Zhu et al. [223] and Caiazzo and Maddu [224] find only first order convergence in space for the velocity field. In section 8.4, it has been observed that the IBM can be of second-order accuracy when the deformation of a capsule in shear flow is considered (additional information also provided in Krüger et al. [187]). One of the most important findings is that the hydrodynamic radius of the particle is larger than expected from the input mesh. The finite-range interpolations of the IBM are believed to be the reason for this behavior. Generally speaking, the accuracy of the IBM seems to depend strongly on the application and observables. Due to the complexity of the mathematical basis of the coupled IBM-LBM system, a thorough analysis of the spatial accuracy is not provided here, and the reader is referred to the literature, e.g., [224].

Even if the Navier-Stokes solver provides a divergence-free velocity field, the velocity field interpolated by IBM will not be divergence-free in general (cf. section 7.4). For this reason, one may track the membrane volumes explicitly and correct for volume drifts, e.g., by introducing a volume restoration force as in eq. (7.15). In the present case, the resulting volume deviations are below 1% and can be neglected.

Le and Zhang [225] reported an unphysical boundary slip velocity in a coupled IBM-LBM system when the relaxation parameter τ is significantly larger than unity. Practically, it should be < 2. Indeed, a similar observation has been pointed out in section 8.4 (and, in more detail, in Krüger et al. [187]). For that reason, the LBM relaxation parameter in the present thesis never exceeds unity except for benchmarking. This is also in line with the restrictions given by the bounce-back boundary conditions in section 5.4.

Restrictions of the membrane model

The membrane consists of flat triangular face elements which remain flat even after deformation. If the local curvature radius becomes comparable to the extension of a face, the mesh resolution is not sufficient any more, and the deformation state is not described in a reliable way. This sets an upper bound for the achievable shear rates which can be simulated. The smaller the membrane resolution (the less faces used), the smaller the shear rate which can be simulated. Moreover, higher particle volume fractions detrimentally decrease the maximum shear rate since membrane-membrane interactions lead to additional deformations. Numerically, too strong a deformation can lead to the collapse of faces or the folding of pairs of faces. Due to the shortcomings of the IBM, such an unphysical deformation usually cannot be reversed, and the simulation becomes either inaccurate or it even crashes[2]. Also Pozrikidis [84] and MacMeccan [194]—using different membrane models—observed that a coarse membrane resolution leads to numerical problems when large deformations are simulated. Local fluid velocities above ≈ 0.1 may lead to instabilities since this translates to a position shift of ≈ 0.1 (about 10% of the extension of a face) in one time step. Unphysical oscillations may therefore emerge and destroy the simulation in the worst case. The maximum velocity restriction ($u < 0.1$) is in line with the small Mach number premise of the LBM.

As explained before, the penalty moduli for the surface and volume conservations, κ_A and κ_V, (cf. sections 7.3 and 7.4) are not allowed to become too large. However, as commonly stated in the literature [31, 77, 79], the exact values of κ_A and κ_V are not relevant as long as the surface and volume are sufficiently well conserved (within 1%, say). It should be noted that the surface and volume deviations become smaller when the spatial resolution is increased. Reasonable values for κ_A and κ_V are not obvious a priori, but they can be inferred from a few test runs.

Due to the discretization of the mesh, the membrane forces may suffer from meshing artifacts which can be caused by extremely small face areas since these enter the bending force in the denominator, cf. appx. C.2. Membrane faces, therefore, should be of comparable size. Additionally, faces with large interior angles are of disadvantage because they cannot accurately capture the shear state of the faces. For this reason, it is desirable to have equisized faces with shapes close to equilateral triangles. This is ensured by the mesh generation algorithm presented in section 8.3. As the connectivity of the mesh is never changed in the present model, nodes cannot 'diffuse' in the mesh, i.e., the mesh has no fluidity properties. Physically, this means that the particles always require a finite *elastic* shear resistance. Vesicles—which are basically 2D fluids—cannot support elastic in-plane shear stresses, and membrane nodes would eventually diffuse. Consequently, the current model supports the simulation of capsules rather than vesicles.

[2]The reason for such a crash is the explicit time integration in combination with large forces which may drive individual membrane nodes out of the numerical grid.

Conclusions

Simulations at arbitrarily small shear rates are only restricted by the available computing time. For high shear rates, the mesh resolution and the lattice Mach number limitation define upper bounds. The present model is particularly suited for small Reynolds number simulations where $\tau \approx 1$ can be selected. The resolutions of the Lagrangian and Eulerian meshes should be similar, $\bar{l} \approx \Delta x$, in order to produce optimum results. High volume fraction simulations may be problematic due to the IBM weakness for small node-node separations. A repulsion force may be incorporated to remove the numerical problem. Yet, the consequences of the repulsion force on the physical results should be monitored. Due to the IBM interpolations, one has to consider potential accuracy issues which have to be checked. The current membrane model should only be applied to capsules (membranes with finite elastic shear resistance) and not to vesicles since the mesh connectivity is fixed.

8.7. Interactions between nearby membranes

It is a well-known problem in lattice computations that the hydrodynamic description breaks down if the distance d between two particles becomes smaller than a length comparable to the lattice constant Δx (e.g., [73, 79, 154, 226]). Even worse, due to the discrete time stepping, particles may eventually overlap if no countermeasures are taken.

For spherical particles, lubrication correction forces can be computed analytically and may be added to re-introduce the correct hydrodynamics at a sub-grid length scale [73, 110]. This approach, however, is difficult to employ in the present model since the particles generally have non-spherical shapes. Additionally, the IBM concept reveals a numerical weakness when particle nodes have a distance much smaller than the lattice resolution Δx (see below). For that reason, a pragmatic method is used in order to maintain a safety distance between the particles: a repulsion force for nodes being too close to each other.

Before details of the appropriated repulsion force are given, a qualitative analysis of the IBM at short distances should be performed. In the framework of the IBM, particle nodes can only translate by moving along with the ambient fluid, and velocity interpolations play an important role, cf. section 6.3. If two particle nodes are close to each other (d significantly smaller than Δx), the nodes basically see the same ambient velocity field, and it becomes more and more difficult to separate the nodes once they have approached each other. In order to understand this, it is instructive to imagine a 1D case with two particle nodes i and j at positions x_i and x_j close to each other, with a mutual distance $x_j - x_i = d < \Delta x$, cf. fig. 8.12. Both nodes are located between two adjacent fluid lattice nodes at positions X_l and X_r with $X_r - X_l = \Delta x$. For this particular case, the velocities of the two nodes can be written as, cf. eq. (6.6),

$$\begin{aligned} \dot{x}_i &= \left(1 - \frac{x_i - X_l}{\Delta x}\right) u(X_l) + \left(1 - \frac{X_r - x_i}{\Delta x}\right) u(X_r), \\ \dot{x}_j &= \left(1 - \frac{x_j - X_l}{\Delta x}\right) u(X_l) + \left(1 - \frac{X_r - x_j}{\Delta x}\right) u(X_r) \end{aligned} \tag{8.12}$$

when ϕ_2, eq. (6.11), is used as interpolation stencil. It is straightforward to show that the relative velocity of the two nodes is

$$\dot{x}_j - \dot{x}_i = \frac{d}{\Delta x} \left(u(X_r) - u(X_l)\right). \tag{8.13}$$

In other words, due to the linear interpolation, the velocity difference which can bring the nodes apart from each other is reduced by a factor $d/\Delta x$ if the particles are close. Although not exactly

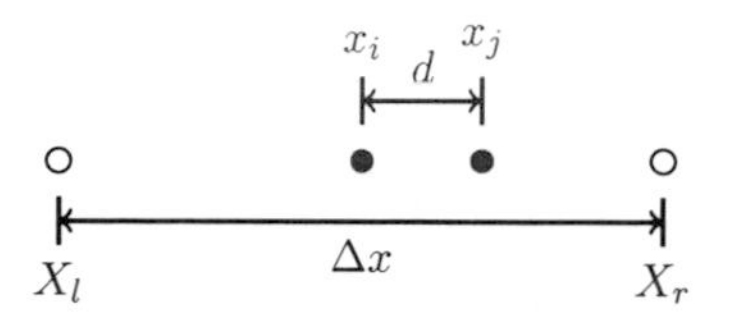

Fig. 8.12.: Immersed boundary method and small node distances. A simplified problem in 1D is shown: Two nodes (red circles) at positions x_i and x_j and mutual distance d are located between two adjacent fluid lattice nodes (white circles) at X_l and X_r with distance Δx.

the same, the results for the other interpolation stencils, ϕ_3 and ϕ_4 in eq. (6.8) and eq. (6.10), are comparable. A qualitatively similar result is also expected in 2D and 3D.

In order to hinder nodes to approach each other too closely, a repulsion force may be employed. It should act along the connection line of the nodes and fulfill $F_j = -F_i$ (in the 1D case) in order to obey momentum and angular momentum conservation. Applied to the above example, it is easy to see that the resulting force acting on lattice node X_l is

$$f(X_l)\Delta x = \left(1 - \frac{x_i - X_l}{\Delta x}\right) F_i + \left(1 - \frac{x_j - X_l}{\Delta x}\right) F_j = \frac{d}{\Delta x} F_i, \tag{8.14}$$

cf. eq. (6.5). The force acting on lattice node X_r is the same with different sign[3]. At this point, the important observation is that the effective repulsion force $f(X_l)\Delta x$ which survives the IBM spreading is reduced by a factor $d/\Delta x$ compared to the input force F_i. For that reason, the repulsion force should increase stronger than linearly when the node distance d goes to zero.

A simple repulsion force is a power-law of the distance d with a cut-off at a finite length R,

$$\boldsymbol{F}_{ij}(\boldsymbol{d}_{ij}) = \begin{cases} -\kappa_{\text{int}} \left[\left(\frac{\Delta x}{d_{ij}}\right)^k - \left(\frac{\Delta x}{R}\right)^k \right] \frac{\boldsymbol{d}_{ij}}{d_{ij}} & \text{for} \quad d_{ij} < R, \\ 0 & \text{for} \quad d_{ij} \geq R \end{cases} \tag{8.15}$$

where $\boldsymbol{F}_{ij} = -\boldsymbol{F}_{ji}$ is the force acting on node i given the distance $\boldsymbol{d}_{ij} = \boldsymbol{x}_j - \boldsymbol{x}_i$ between node i and a nearby node j. The power k has to be larger than unity in order to overcome the interpolation artifact in eq. (8.14). The magnitude of the repulsion is controlled via the parameter κ_{int}. The unphysical parameters k, R, and κ_{int} have to be chosen in such a way that (i) the force is only active at distances as small as possible, that (ii) the separation of nodes does not become too small ($0.5\Delta x$, say), and that (iii) the repulsion force does not lead to numerical instabilities. In the present thesis, $k = 2$ and $R = \Delta x$ are used. Feng and Michaelides [79], following a similar idea, state that details of the repulsive force do not influence the macroscopic behavior.

It has to be stated that a repulsive force increases the apparent radius of the particles if the particles are in close contact. On the one hand, as long as the particles have a distance larger than the interaction range, the interaction force is deactivated. On the other hand, for dense suspensions, particles are often in close proximity, and the repulsive force is acting. Still, the range of this force is comparable to the additional effective radius caused by the IBM interpolations (cf. section 8.4). The interaction force as defined in eq. (8.15), therefore, can be considered a 'natural' force avoiding overlap of the hydrodynamic particle volumes. It is an open question up to which volume fraction this interpretation is still valid. For large volume fractions, particles have to deform even in the absence of flow. However, in a quiescent situation, hydrodynamic effects are absent whereas the repulsion forces are still active.

It is important to stress that—in the present model—only nodes in different membranes interact via this kind of repulsion force. By design, the average distance of neighboring membrane nodes

[3]This must be the case since the IBM interpolation stencils obey momentum conservation.

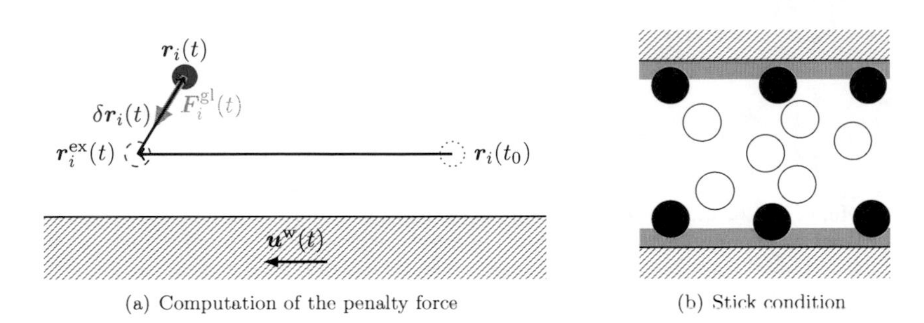

(a) Computation of the penalty force (b) Stick condition

Fig. 8.13.: Implementation of wall roughness. (a) A membrane node i initially located at position $\boldsymbol{r}_i(t_0)$ (dotted circle) close to a wall (dashed region) is expected to move along with the wall to position $\boldsymbol{r}_i^{\text{ex}}(t)$ (dashed circle) after time $t - t_0$, but it is generally found at position $\boldsymbol{r}_i(t)$ (solid circle) instead. The stick force $\boldsymbol{F}_i^{\text{gl}}(t)$ is proportional to the distance $\delta\boldsymbol{r}_i(t) = \boldsymbol{r}_i(t) - \boldsymbol{r}_i^{\text{ex}}(t)$, cf. eq. (8.17). (b) After growing the particles to their full size, all particles which have at least one mesh node in the 'glue' region (light blue) near the walls (dashed regions) are stuck to the wall. Stuck particles are shown as black, free particles as white circles.

in the undeformed state is close to Δx, cf. section 8.3. Even for strong deformations, nodes in a membrane usually do not come closer than half a lattice spacing. This behavior is caused by the in-plane tensions related to the incompressibility constraints of the membrane, cf. section 7.1. The repulsion force, therefore, is only switched on when the distance between two distinct membranes falls below the range R, which is virtually never the case if the suspension is dilute. In these cases, hydrodynamic interactions usually maintain a given distance between the membranes. If the suspensions become denser and denser, the mutual distances between membranes decrease, and, eventually, numerical problems would arise without a repulsion force.

Nodes which come too close to a wall are subject to an additional force introducing a repulsion away from the wall. The functional form is the same as in eq. (8.15) with the difference that the distance $\boldsymbol{d}_{ij}$ is replaced by the shortest (i.e., normal) distance to the wall. Since the wall is not allowed to move in normal direction, it is not necessary to add the repulsion force to the wall. Such a force would merely increase the wall pressure which is not of significance in the present thesis.

8.8. Wall slip and roughness

In experiments, the rheological properties of a suspension strongly depend on the structure of the shearing surfaces. It is known that smooth surfaces promote a slip behavior which reduces the viscosity of the suspension at small shear stresses [227, 228]. The reason is a thin fluid lubrication film developing between the surface and the first layer of suspended particles. If the shear stress is close to or smaller than the yield stress of the suspension, the entire velocity gradient can drop off in the fluid film, resulting in locally high shear rates. Even though the suspension in the bulk is jammed, the overall suspension can flow, and a yield stress may be hidden from the observer. If slip effects are not desired, one may either subtract the slip velocity (e.g., by measuring the velocity profile locally) or prevent slippage in the first place.

The development of the lubrication layer can be avoided if rough (on the particle length scale) surfaces are used. For the present model, the wall roughness may be realized in the following way: The particles are randomly initialized in the available volume as described in section 8.5. After this, all particles in a given vicinity of the wall are 'stuck to the wall'. In order to do so, the nodes of these membranes experience an additional force trying to keep the nodes in a position where they would be located if they moved along with the wall. This concept is visualized in fig. 8.13(a). At the beginning of the simulation, after the membranes have grown to their full size and membranes close to the walls have been identified, each node of these membranes remembers its initial position $\boldsymbol{r}_i(t_0)$. At a later time, $t > t_0$, the wall has moved by a certain distance and with it the expected node position $\boldsymbol{r}_i^{\mathrm{ex}}(t)$ in such a way that

$$\boldsymbol{r}_i^{\mathrm{ex}}(t) = \boldsymbol{r}_i(t_0) + \int_{t_0}^{t} \mathrm{d}t' \, \boldsymbol{u}^{\mathrm{w}}(t') \tag{8.16}$$

where $\boldsymbol{u}^{\mathrm{w}}$ is the wall velocity. Generally, the node will be at a position $\boldsymbol{r}_i(t)$ instead, with relative distance $\delta\boldsymbol{r}_i(t) = \boldsymbol{r}_i(t) - \boldsymbol{r}_i^{\mathrm{ex}}(t)$ to its expected position. A Hookean penalty force for this node is defined according to

$$\boldsymbol{F}_i^{\mathrm{gl}}(t) = -\kappa_{\mathrm{gl}} \delta\boldsymbol{r}_i(t). \tag{8.17}$$

The strength of the force is given by the 'glue modulus' κ_{gl}. This force is added to the other forces acting on the nodes. An opposite force with the same magnitude is exerted on the corresponding wall in order to find the correct wall shear stress afterwards. This way, momentum conservation is not violated.

In practice, a membrane is stuck to a wall if at least one of its nodes has a wall distance not greater than the large radius r of the membranes. However, other conventions may be used as well. This can be interpreted as a glue film of thickness r covering both walls, cf. fig. 8.13(b). The above method bases on the approach presented by Feng and Michaelides [79] who use a Hookean penalty force to maintain the shape of the pseudo-rigid particles.

Part III.

Simulation results and interpretation

9. Stress evaluation in combined immersed boundary lattice Boltzmann simulations

The recovery of the stress tensor in computer simulations is not as straight-forward as it may appear at first view. The reason is that usually not the stress tensor $\boldsymbol{\sigma}$ itself enters the macroscopic equations. Rather, its divergence, $\nabla \cdot \boldsymbol{\sigma}$ appears. Even if this divergence is known, it is generally not possible to uniquely reconstruct the stress tensor from it since the equation system is under-determined[1]. It has already been mentioned in section 5.2 that the full fluid stress tensor is known at each point within the lattice Boltzmann method (LBM). The situation is different for the particle stress. If, however, averages of some kind are sufficient (e.g., over time, volume, or a coordinate plane), different approaches are available to recover the particle stress tensor or some of its components.

In this chapter, it is argued how suspension stresses can be evaluated within the model presented in the previous chapters. The discussion is limited to viscous flows where inertia effects are negligible. The fluid stress considerations are briefly summarized in section 9.1, followed by the wall stress in section 9.2. The direct recovery of particle stresses is more demanding as discussed in sections 9.3 and 9.4. Verification simulations are presented in section 9.5, linking the above approaches.

9.1. Fluid stress evaluation in the lattice Boltzmann method

Simple fluids cannot support elastic stresses, and the fluid stress equals the viscous stress [2],

$$\boldsymbol{\sigma}^{\mathrm{f}} = 2\eta_0 \boldsymbol{S}, \tag{9.1}$$

where $S_{\alpha\beta} = \frac{1}{2}(\partial_\alpha u_\beta + \partial_\beta u_\alpha)$, cf. eq. (2.5), are the components of the symmetric fluid shear rate tensor and η_0 is the dynamic shear viscosity. Bulk stresses due to compressibility effects are neglected here and in the following.

For a Newtonian fluid, the dynamic shear viscosity does not depend on the shear rate, and the stress is proportional to the shear rate. Water and blood plasma are examples of Newtonian fluids over a wide range of shear rates, including physiological shear rates up to about $10^4\,\mathrm{s}^{-1}$. The fluid described in the standard LBM is Newtonian as long as the relaxation parameter τ is a constant. Although LBM extensions for non-Newtonian fluids exist (e.g., [229, 230, 231]), the suspending fluid is always Newtonian in the present work. Non-Newtonian suspension rheology emerges from the presence of particles immersed in the fluid.

For conventional Navier-Stokes solvers where the Navier-Stokes equations (NSE) are directly discretized, the fluid stress is usually computed from the velocity field via differentiation. In lattice Boltzmann simulations, however, the full fluid stress tensor is accessible at each point and without evaluating velocity gradients, cf. eq. (5.17). This makes the LBM an attractive Navier-Stokes solver when the rheology of fluids is to be investigated.

[1] The divergence $\nabla \cdot \boldsymbol{\sigma}$ yields three equations, but six independent components of the symmetric stress tensor $\boldsymbol{\sigma}$ are required.

9.2. Wall stress evaluation in the lattice Boltzmann method

The wall stress tensor $\boldsymbol{\sigma}^{\mathrm{w}}$ is related to the total force $\Delta \boldsymbol{p}/\Delta t$ acting on a small patch of the oriented wall surface $\Delta \boldsymbol{A}$ in a given time Δt as defined in eq. (5.28),

$$\frac{\Delta \boldsymbol{p}}{\Delta t} = \boldsymbol{\sigma}^{\mathrm{w}} \cdot \Delta \boldsymbol{A}, \tag{9.2}$$

where $\Delta \boldsymbol{p}$ is the corresponding momentum exchange in time Δt. In the present work, only the shear components are considered. They are caused by the force components parallel to the wall, $\Delta \boldsymbol{p}_{\parallel} \cdot \Delta \boldsymbol{A} = 0$.

For a suspension, the wall stress is the sum of the fluid stress and the particle stress at the wall. Due to lubrication effects and hydrodynamic lift forces, particles are usually in no direct contact with a wall [227], and $\Delta \boldsymbol{p}/\Delta t$ is the force due to the momentum exchange of the fluid caused by the no-slip boundary condition. Within the LBM, the wall stress can be evaluated at each surface patch $\Delta \boldsymbol{A}$ from eq. (5.27).

If the walls are made rough as discussed in section 8.8, the stick forces have to be considered in the computation of the wall stress, and the total force acting on the entire wall is the sum of the fluid forces and the negative of the stick forces ('actio = reactio') as defined in eq. (8.17).

For steady and simple shear flow, the condition of mechanical stability demands that the wall shear stresses at the bottom and top walls are identical and equal to the shear stress everywhere else in the system. For this reason, an evaluation of the wall stress is in principle sufficient if the total suspension stress and the average viscosity are to be computed. This is the approach commonly followed in rheology experiments. However, it does not allow the local separation of fluid and particle stresses, and, therefore, the local viscosity between the walls cannot be accessed. In some cases, the local stresses are of high relevance, especially if wall effects are important and the system is not homogeneous [5]. It is desirable to measure the contributions of the fluid and the particles to the total stress locally and *independently*. In experiments, a local stress measurement is extremely difficult if not impossible. Even in simulations, it is not a priori clear how to evaluate local particle stresses. Two possible approaches are presented in the following sections.

9.3. Evaluating particle stresses with Batchelor's approach

In a simple fluid, the stress $\boldsymbol{\sigma}$ is of viscous nature only, and one can write $\boldsymbol{\sigma} = 2\eta_0 \boldsymbol{S}$ at each point in the fluid. If particles—deformable or not—are suspended in this fluid, there are additional stress contributions caused by the distorted velocity field due to the presence of the particles. In the following, only the shear component (xz-component) of the stress in simple shear flows (velocity along x-axis, velocity gradient along z-axis) is considered. All other stress components are either not relevant for the present discussion, or they vanish on average. The apparent viscosity is defined via the volume average

$$\langle \sigma_{xz} \rangle_V = 2\eta_{\mathrm{app}} \langle S_{xz} \rangle_V. \tag{9.3}$$

Suspension stress

It is possible to compute the apparent viscosity for a dilute suspension of rigid, spherical particles in simple shear flow [2, 14]. The Einstein relation states that, for a volume fraction ϕ not larger than a few percent, the apparent viscosity is $\eta_{\mathrm{app}} = \eta_0 \left(1 + \frac{5}{2}\phi\right)$, cf. section 2.3. For arbitrary volume fractions and general particle shapes and deformabilities, it is either extremely difficult or even impossible to find the apparent viscosity analytically. Instead, numerical approaches may be required.

In his seminal work, Batchelor [10] derived a general, formal expression for the stress in suspensions subject to shear flow. Starting from the NSE, Batchelor first introduced the *bulk stress* including pressure,

$$\boldsymbol{\Sigma} := \langle -p\boldsymbol{I} + \boldsymbol{\sigma} \rangle_V = \frac{1}{V} \int \mathrm{d}V \, (-p\boldsymbol{I} + \boldsymbol{\sigma}), \tag{9.4}$$

as the volume average of the local stress. At this point, locality is already lost. It has been further shown that the bulk stress can be written in the form

$$\boldsymbol{\Sigma} = -\frac{1}{V} \int_{\text{fluid}} \mathrm{d}V \, p\boldsymbol{I} + 2\eta_0 \langle \boldsymbol{S} \rangle_V + \langle \boldsymbol{\sigma}^{\mathrm{P}} \rangle_V \tag{9.5}$$

where the pressure is only integrated over the fluid volume (volume not occupied by particles). This isotropic pressure contribution is not of interest here, and it is neglected in the following. As can be inferred from eq. (9.5), the total stress can be written as the sum of the *known* fluid stress $2\eta_0 \langle \boldsymbol{S} \rangle_V$ as it would be in the absence of the particles and the particle stress $\langle \boldsymbol{\sigma}^{\mathrm{P}} \rangle_V$. Exploiting Gauss' theorem, the particle stress can be shown to have the components

$$\langle \sigma^{\mathrm{P}}_{\alpha\beta} \rangle_V = \frac{1}{V} \sum_k \oint_{A_k} \mathrm{d}A \, \left(S_{\alpha\gamma} x_\beta n_\gamma - \eta_0 (u_\alpha n_\beta + u_\beta n_\alpha) \right). \tag{9.6}$$

The sum runs over all suspended particles, and the integration is taken over particle surfaces with the unit normal vector $\boldsymbol{n}$ pointing into the fluid[2]. $\boldsymbol{x}$ is the position vector with an arbitrary origin. Eq. (9.5) and eq. (9.6) are generally valid for negligible inertia effects and for a Newtonian suspending fluid at any instance of time. The particle shape and the volume fraction ϕ are not restricted in any form.

Application to immersed elastic membranes

In the following, the above formalism will be applied to thin membranes immersed in a fluid. These membranes are filled with another Newtonian fluid of viscosity $\lambda\eta_0$ where λ is the viscosity ratio. In this case, both the exterior and the interior surfaces have to be considered, and one can write

$$\begin{aligned} \langle \sigma^{\mathrm{P}}_{\alpha\beta} \rangle_V = & \frac{1}{V} \sum_k \oint_{A_k^+} \mathrm{d}A \, \left(S_{\alpha\gamma} x_\beta n^+_\gamma - \eta_0 (u_\alpha n^+_\beta + u_\beta n^+_\alpha) \right) \\ & + \frac{1}{V} \sum_k \oint_{A_k^-} \mathrm{d}A \, \left(S_{\alpha\gamma} x_\beta n^-_\gamma - \lambda\eta_0 (u_\alpha n^-_\beta + u_\beta n^-_\alpha) \right). \end{aligned} \tag{9.7}$$

Each membrane has an exterior and an interior surface which are denoted by + and −, respectively. Thus, A^+ lies in the exterior fluid with viscosity η_0 and A^- in the interior fluid with viscosity $\lambda\eta_0$. Due to the small thickness of the membrane, both surfaces have an infinitesimal distance ϵ from each other, and the normal vectors $\boldsymbol{n}^+$ and $\boldsymbol{n}^-$ pointing into the exterior and the interior fluid obey $\boldsymbol{n}^+ = -\boldsymbol{n}^-$. Since the velocity $\boldsymbol{u}$ is smooth at the membrane surface (no-slip condition), the particle stress can be written in the form

$$\langle \sigma^{\mathrm{P}}_{\alpha\beta} \rangle_V = \frac{1}{V} \sum_k \left(\oint_{A_k^+} - \oint_{A_k^-} \right) \mathrm{d}A \, S_{\alpha\gamma} x_\beta n_\gamma + \frac{1}{V} \sum_k \oint_{A_k} \mathrm{d}A \, (\lambda - 1)\eta_0 (u_\alpha n_\beta + u_\beta n_\alpha) \tag{9.8}$$

[2] The integration is performed in the exterior fluid, directly outside of the suspended particles where fluid velocity and shear rate are defined.

where $\boldsymbol{n} = \boldsymbol{n}_+ = -\boldsymbol{n}_-$ has been substituted. Tensions in the membrane are balanced by a jump of the fluid stress across the interface [129], and

$$\tilde{\boldsymbol{f}} = (\boldsymbol{S}^- - \boldsymbol{S}^+) \cdot \boldsymbol{n} \tag{9.9}$$

holds where $\boldsymbol{S}^+$ and $\boldsymbol{S}^-$ are the values of the fluid stress tensor directly outside and inside of the membrane, respectively. The force density $\tilde{\boldsymbol{f}}$ (force per area) is exerted *on the fluid* by the membrane[3], and for a given membrane deformation it is known from the constitutive model, cf. chap. 7. Consequently, the particle stress is [25]

$$\langle \sigma^{\mathrm{P}}_{\alpha\beta} \rangle_V = \frac{1}{V} \sum_k \oint_{A_k} \mathrm{d}A \left(-\tilde{f}_\alpha x_\beta + (\lambda - 1)\eta_0 (u_\alpha n_\beta + u_\beta n_\alpha) \right). \tag{9.10}$$

In the present work, the interior fluid has the same viscosity as the exterior fluid, $\lambda = 1$, and the bulk particle stress reduces to the compact form

$$\langle \sigma^{\mathrm{P}}_{\alpha\beta} \rangle_V = -\frac{1}{V} \sum_k \oint_{A_k} \mathrm{d}A \, \tilde{f}_\alpha x_\beta. \tag{9.11}$$

Remarks

It is straightforward to evaluate eq. (9.11) within the present model. The discretization of eq. (9.11) reads

$$\langle \sigma^{\mathrm{P}}_{\alpha\beta} \rangle_V = -\frac{1}{V} \sum_i F_{i\alpha} x_{i\beta} \tag{9.12}$$

where the sum runs over all Lagrangian membrane nodes i (force $\boldsymbol{F}_i$, position $\boldsymbol{x}_i$) in the entire simulation box. The origin of the coordinate system is arbitrary if the sum of all forces in the system is exactly zero, $\sum_i \boldsymbol{F}_i = 0$. Indeed, the definition of the particle stress in eq. (9.11) is only useful if there is no net force on the particles. Else, the particle stress could take any value by choosing a convenient coordinate origin.

The particle stress as given in eq. (9.11) is the *average* particle stress in the entire system. There is a priori no access to a local particle stress within Batchelor's approach. However, it is possible to compute the contributions of individual particles k,

$$\langle \sigma^{\mathrm{P}}_{k\alpha\beta} \rangle_V := -\frac{1}{V} \oint_{A_k} \mathrm{d}A \, \tilde{f}_\alpha x_\beta. \tag{9.13}$$

This stress may then be considered as being located at the centroid of the particle. It will be shown in section 9.5 that this approach does not give satisfactory results. Moreover, interacting particles lead to problems since for such a system, the net force on an individual particle is not zero in general. Interaction forces should be excluded from eq. (9.13).

9.4. Evaluating local particle stresses with the method of planes

The approach presented in section 9.3 allows to obtain the volume average of the particle stress, $\langle \sigma^{\mathrm{P}}_{\alpha\beta} \rangle_V$. While the fluid stress $\boldsymbol{\sigma}^{\mathrm{f}}$ can be obtained locally (section 9.1), this is not possible for the particle stress up to this point. In principle, the particle stress may be obtained indirectly. For example, it is known that, for a simple shear flow, the shear stress averaged over time

[3] An opposite force is exerted on the membrane by the fluid, which is the reason for contradicting sign conventions often found in the literature.

and the xy-plane is constant throughout the system, $\langle \sigma_{xz} \rangle_{x,y,t} \neq \langle \sigma_{xz} \rangle_{x,y,t}(z)$. The fluid stress $\langle \sigma^{\mathrm{f}}_{xz} \rangle_{x,y,t}(z)$ is known, so the particle stress is

$$\langle \sigma^{\mathrm{p}}_{xz} \rangle_{x,y,t}(z) = \langle \sigma_{xz} \rangle_{x,y,t} - \langle \sigma^{\mathrm{f}}_{xz} \rangle_{x,y,t}(z). \tag{9.14}$$

For a rheological study, this relation may in principle be sufficient. Still, within this approach, there is no access to spatio-temporal fluctuations of the particle stress. These fluctuations carry important additional information about the system, e.g., an independent measure of shear viscosity in the small shear rate regime [232],

$$\eta \propto \int_0^\infty \mathrm{d}t' \, \langle \sigma_{xz}(0) \sigma_{xz}(t') \rangle_V. \tag{9.15}$$

In this section, another approach for the particle stress evaluation is presented (Krüger et al. [233]). It offers the possibility to find the *instantaneous and local* particle stress on a plane parallel to the confining walls. For a special case, this technique is shown to be identical to the 'method of planes' (MOP) which has been introduced by Todd et al. [4] for the case of a simple liquid and further examined by Varnik et al. [5] in the case of a polymer melt.

Without external forces, the NSE can be written as

$$\boldsymbol{g}(\boldsymbol{r}, t) = \nabla \cdot \boldsymbol{\sigma}(\boldsymbol{r}, t) \tag{9.16}$$

where $\boldsymbol{g}$ contains the convective derivative of the velocity and the pressure gradient. In the present model, however, the lattice Boltzmann stress tensor only captures the fluid component, whereas the particle contribution is contained in the force density $\boldsymbol{f}$,

$$\boldsymbol{g}(\boldsymbol{r}, t) = \nabla \cdot \boldsymbol{\sigma}^{\mathrm{f}}(\boldsymbol{r}, t) + \boldsymbol{f}(\boldsymbol{r}, t). \tag{9.17}$$

Therefore, the first step is to assume that the particle stress and the membrane force density are connected via

$$\boldsymbol{f}(\boldsymbol{r}, t) = \nabla \cdot \boldsymbol{\sigma}^{\mathrm{p}}(\boldsymbol{r}, t). \tag{9.18}$$

This fundamental relation is local both in space and time and known to be valid for elastic systems in equilibrium, i.e., in the absence of accelerations [189]. Eq. (9.18) states that the effect of interactions on flow behavior can be incorporated in the NSE either by (i) direct implementation of particle forces as a (spatially and temporally varying) external force field or (ii) by introducing the particle stress tensor. For any differentiable stress field $\boldsymbol{\sigma}^{\mathrm{p}}(\boldsymbol{r}, t)$, a corresponding force density $\boldsymbol{f}(\boldsymbol{r}, t)$ can be obtained.

Stress evaluation

The α-component of eq. (9.18) can be written as

$$f_\alpha(\boldsymbol{r}, t) = \partial_x \sigma^{\mathrm{p}}_{\alpha x}(\boldsymbol{r}, t) + \partial_y \sigma^{\mathrm{p}}_{\alpha y}(\boldsymbol{r}, t) + \partial_z \sigma^{\mathrm{p}}_{\alpha z}(\boldsymbol{r}, t). \tag{9.19}$$

For periodic boundary conditions along the x- and y-axes, as in the present work, and averaging over the xy-plane, this equation simplifies to

$$\langle f_\alpha \rangle_{x,y}(z, t) = \partial_z \langle \sigma^{\mathrm{p}}_{\alpha z} \rangle_{x,y}(z, t). \tag{9.20}$$

Integration yields

$$\langle \sigma^{\mathrm{p}}_{\alpha z} \rangle_{x,y}(z, t) = \langle \sigma^{\mathrm{p}}_{\alpha z} \rangle_{x,y}(z_0, t) + \int_{z_0}^{z} \mathrm{d}z' \, \langle f_\alpha \rangle_{x,y}(z', t) \tag{9.21}$$

which is a very important intermediate result. It states that, if the force density $\boldsymbol{f}(\boldsymbol{r},t)$ and the xy-average of the particle stress at position z_0 are known, then the xy-average of the particle stress at each z-position is known.

In its discretized form, the force density in the Lagrangian system can be written as

$$\boldsymbol{f}(\boldsymbol{r},t) = \sum_i \boldsymbol{F}_i(t)\delta(\boldsymbol{r} - \boldsymbol{x}_i(t)) \tag{9.22}$$

where $\boldsymbol{F}_i$ is the force acting on Lagrangian node i which is located at point $\boldsymbol{x}_i(t)$ at time t. Interestingly, the particle membership of node i and the physical origin of $\boldsymbol{F}_i$ do not play a role. In particular, it is not necessary to claim that $\boldsymbol{F}_i$ is a two-body force. This is important for multi-body forces which enter $\boldsymbol{F}_i$, e.g., through bending or volume contributions. The discretized form of the particle stress reads

$$\langle\sigma^{\mathrm{P}}_{\alpha z}\rangle_{x,y}(z,t) = \langle\sigma^{\mathrm{P}}_{\alpha z}\rangle_{x,y}(z_0,t) + \frac{1}{A}\sum_i F_{i\alpha}(t)\theta(z_i(t) - z_0)\theta(z - z_i(t)) \tag{9.23}$$

where $A = L_x L_y$ (L_x and L_y being the system extensions along the x- and y-directions) and $\theta(z)$ is the Heaviside step function, i.e., all Lagrangian nodes between z_0 and z contribute.

Method of planes

It will be shown in the following that eq. (9.23) reduces to the equation proposed by Todd et al. [4],

$$\langle\sigma^{\mathrm{P}}_{\alpha z}\rangle_{x,y}(z,t) = -\frac{1}{2A}\sum_i F_{i\alpha}(t)\operatorname{sgn}(z_i(t) - z), \tag{9.24}$$

for some additional assumptions. Here, $\operatorname{sgn}(z)$ is the sign function. The first assumption is that the particle stress vanishes at z_0, $\langle\sigma^{\mathrm{P}}_{\alpha z}\rangle_{x,y}(z_0,t) = 0$, and that no nodes exist with $z_i(t) < z_0$. Thus, z_0 may be taken as the position of the impenetrable bottom wall. The second claim is that the total force vanishes, $\sum_i F_{i\alpha}(t) = 0$, i.e., the total momentum is conserved. This translates to $\sum_i F_{i\alpha}(t)\theta(z - z_i(t)) = -\sum_i F_{i\alpha}(t)\theta(z_i(t) - z)$ and

$$\begin{aligned}\sum_i F_{i\alpha}(t)\theta(z - z_i(t)) &= \frac{1}{2}\left(\sum_i F_{i\alpha}(t)\theta(z - z_i(t)) - \sum_i F_{i\alpha}(t)\theta(z_i(t) - z)\right) \\ &= \frac{1}{2}\sum_i F_{i\alpha}(t)\operatorname{sgn}(z - z_i(t)).\end{aligned} \tag{9.25}$$

Hence, the formal connection between eq. (9.23) and eq. (9.24) has been shown.

Remarks

The MOP [4] has originally been introduced to find the local stress in an atomistic non-equilibrium fluid in planar geometries. Interestingly, it can be applied directly to the present problem by assuming that the deformable membranes consist of interacting point particles. Formally, this is actually the case because the forces acting on the membranes are considered as being concentrated at the Lagrangian nodes. The physical origin of these forces do not play a role at all.

It must be emphasized again that stresses due to inertia effects are not considered here. On the one hand, the fluid inertia is neglected in the NSE in the first place. On the other hand, the equilibrium condition, eq. (9.18), in the absence of accelerations is considered. As will be shown in section 9.5, the MOP produces excellent results for the local particle stress.

The MOP as presented in eq. (9.23) bases on the forces defined in the Lagrangian system. It is also possible to compute the particle stresses in the Eulerian frame by using the lattice force density as obtained from the immersed boundary spreading via eq. (6.5), i.e., the delta function in eq. (9.22) is replaced by the smooth interpolation stencil, e.g., eq. (6.11). For the present model, this second approach will be employed if not otherwise stated. The reason is that the fluid is driven by the forces in the Eulerian system. It is not aware of the presence of the membranes otherwise. This way, both the fluid and the particle stresses are computed in the Eulerian system. Eq. (9.23) then becomes

$$\langle\sigma^{\mathrm{p}}_{\alpha z}\rangle_{x,y}(z,t) = \langle\sigma^{\mathrm{p}}_{\alpha z}\rangle_{x,y}(z_0,t) + \frac{1}{A}\sum_{\boldsymbol{X}'} f_\alpha(\boldsymbol{X}',t)\Delta x^3\theta(z'-z_0)\theta(z-z') \tag{9.26}$$

where the sum runs over all lattice nodes at position $\boldsymbol{X}' = (x', y', z')$ with force density $\boldsymbol{f}(\boldsymbol{X}',t)$, i.e., the force acting on one lattice unit volume Δx^3 is $\boldsymbol{f}(\boldsymbol{X}',t)\Delta x^3$ (Krüger et al. [233]).

9.5. Benchmark test: verification of the stress evaluation methods

Two classes of benchmark tests have been performed to show (i) the reliability and capability of the stress evaluation methods described in this chapter, (ii) the wall roughening procedure (section 8.8), and (iii) the shear stress boundary condition (section 5.4.2). A single spherical capsule in simple shear flow is considered in section 9.5.1, whereas a dense suspension of spherical capsules (50% volume fraction) is simulated in section 9.5.2. All quantities are given in lattice units.

9.5.1. Single capsule in shear flow

A single spherical capsule ($r = 8$, $N_f = 2000$, $\kappa_{\mathrm{S}} = 0.1$, $\kappa_\alpha = 1$, $\kappa_{\mathrm{B}} = 0.01$, $\kappa_{\mathrm{A}} = 1$, $\kappa_{\mathrm{V}} = 1$) is placed in the middle of an initially quiescent fluid (volume $30 \times 30 \times 30$) bounded by two rigid walls at $z = 0$ and $z = 30$. The LBM relaxation parameter is $\tau = 1$, and the average fluid density is unity. Two simulations for a single particle in shear flow have been performed, one with velocity boundary conditions (VBC), the other with shear stress boundary conditions (SBC).

In the first simulation, both walls are moved along the x-axis in opposite directions with a constant velocity of ± 0.02, resulting in an average fluid shear rate $\langle\dot{\gamma}\rangle_V = 1.33 \times 10^{-3}$. The average fluid stress, therefore, is $\langle\sigma^{\mathrm{f}}_{xz}\rangle_V = 2.22 \times 10^{-4}$, independent of the velocity profile between the walls. The results for the particle stress obtained from the wall stress and Batchelor's approach are shown as function of time in fig. 9.1(a). After an initial transient in which the system is not in steady state, both stresses become equal. The time-averaged particle stress between $t = 4000$ and $t = 10000$ is $\langle\sigma^{\mathrm{p}}_{xz}\rangle_{V,t} = 5.67 \times 10^{-5}$ in both cases. For the average stresses obtained from the MOP, there are slightly different results depending on whether the stress is evaluated in the Eulerian or the Lagrangian frame. In the former, it equals the value obtained before, $\langle\sigma^{\mathrm{p}}_{xz}\rangle_{V,t} = 5.67 \times 10^{-5}$. In the latter, it is slightly larger, $\langle\sigma^{\mathrm{p}}_{xz}\rangle_{V,t} = 5.73 \times 10^{-5}$. The reason for this deviation is caused by the immersed boundary force spreading from the Lagrangian to the Eulerian system. Still, the deviation is only 1%. The curve of the time evolution of the volume-averaged particle stress obtained by the MOP in the Eulerian frame collapses with the curve for Batchelor's stress. For this reason, the MOP stress is not shown separately. This observation is a strong indication for the reliability and consistency of the stress evaluation approaches. The fluid stress, evaluated independently and averaged over time (between $t = 4000$ and 10000) and the total volume, is $\langle\sigma^{\mathrm{f}}_{xz}\rangle_{V,t} = 2.22 \times 10^{-4}$ as expected.

For the second simulation, the walls are subject to the SBC. The prescribed shear stress is 2.79×10^{-4} which is the sum of the fluid and particle stresses obtained from the previous

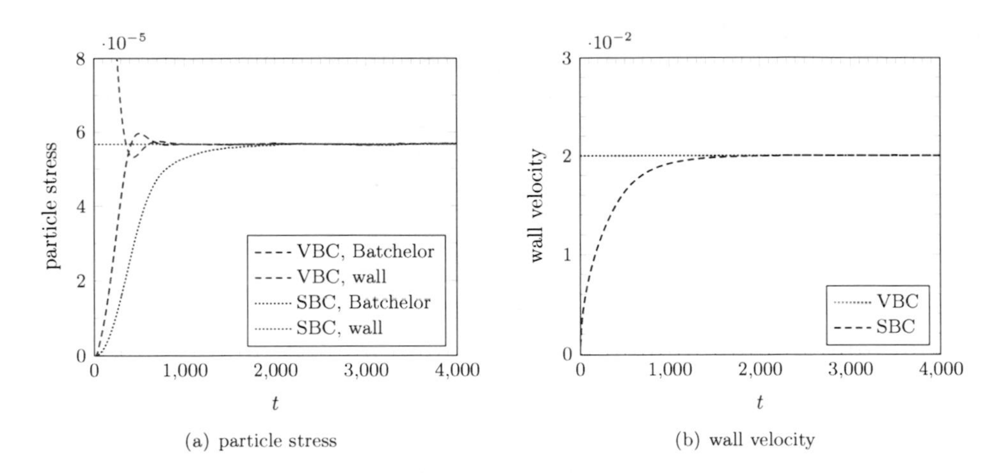

(a) particle stress

(b) wall velocity

Fig. 9.1.: Time evolution of particle stress and wall velocity for a single capsule. The results for the velocity boundary condition (VBC) and the shear stress boundary condition (SBC) are compared. The average particle stress $\langle \sigma^{\mathrm{P}}_{xz} \rangle_V$ has been evaluated using (i) Batchelor's approach and (ii) the wall stress (minus the average fluid stress). In steady state, all results are identical. For the VBC, the wall is impulsively accelerated from zero velocity to 0.02, and the resulting wall stress is initially large. For the SBC, the wall is continuously accelerated, and the transient is longer. The slight fluctuations in steady state are caused by the discrete particle mesh.

simulation run in steady state. All other simulation parameters are then same. The most important observation is that, after a transient, both wall velocities become constant with $u^{\mathrm{w}} = 0.02$ as in the first simulation, cf. fig. 9.1(b). All other steady-state stresses are found to be equal, cf. fig. 9.1(a) and fig. 9.2(a). Again, the time evolutions of Batchelor's stress and the volume-averaged MOP stress are identical (MOP stress not shown separately).

Conclusions and remarks

The above discussion clearly shows the consistency of the SBC for a single particle in simple shear flow. After the initial transients, both simulations are equivalent (fig. 9.1 and fig. 9.2). There are some minor fluctuating deviations between the VBC and the SBC which are caused by the discrete particle mesh.

Due to the observed deviations of the stress results obtained from the MOP in the Eulerian and the Lagrangian frames, only results computed in the Eulerian frame will be presented in the remainder of this thesis. This is more consistent since the presence of the membranes is felt by the fluid only through the forces in the *Eulerian* frame.

Fig. 9.2(a) nicely shows that the local total stress (sum of fluid and particle stresses) is indeed not a function of the transverse coordinate z. Additionally, its value equals these obtained from the wall stress and Batchelor's approach. This is convincing evidence that the MOP actually provides access to the local particle stress *independently* from assumptions based on macroscopic considerations. Without the MOP, there would be no access to the local particle stress[4]. Instead, one may use Batchelor's approach to find the integrated particle stress for the capsule and localize it at the capsule's centroid position ($z = 15$). This, however, would lead to a single data point in the middle of the flow, and the total stress would not be constant (Krüger et al. [233]).

The local viscosity of the suspension is shown in fig. 9.2(b). For regions filled with fluid only, the

[4]Local means stress averaged over the xy-plane as function of z.

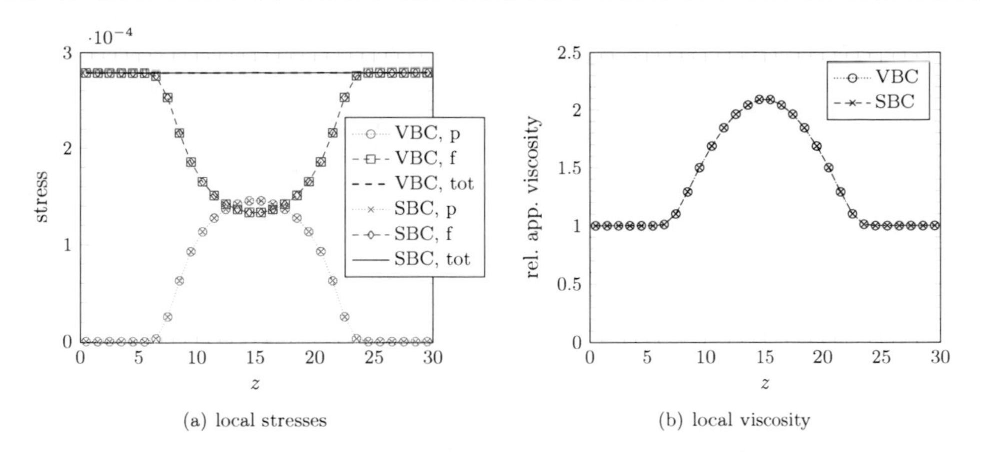

(a) local stresses

(b) local viscosity

Fig. 9.2.: Stresses and viscosity for a single capsule. (a) The local, time-averaged stress contributions in steady state are shown for the velocity boundary condition (VBC) and the shear stress boundary condition (SBC). The particle stresses are computed with the MOP in the Eulerian frame. The total stress (sum of particle and fluid stresses) is a constant line, reflecting the condition of mechanical stability. (b) The local, time-averaged relative apparent viscosity is shown. In the fluid region, it is unity as expected. The presence of the particle in the center region (between $z = 7$ and $z = 23$) gives rise to a finite particle stress and increases the viscosity. Curves for VBC and SBC cannot be distinguished in (a) and (b). All time averages are taken between $t = 4000$ and $t = 10000$.

viscosity equals the suspending fluid viscosity. The presence of the particle, however, increases the local viscosity.

9.5.2. Dense suspension in shear flow

In this section, the capability of the wall roughening (section 8.8) in combination with the stress evaluation is demonstrated. Four simulations have been performed: two with rough and two with smooth walls (VBC and SBC each). The average volume fraction in the simulations is 0.5 (189 spherical particles with an average radius of 5 and a polydispersity of 20%). Depending on the radius of the particles, different meshes have been used. This way, it is guaranteed that the average edge length $\bar{l}$ is as close as possible to the lattice constant Δx. In total, six different meshes have been employed, the smallest with 320, the largest with 1620 faces. The system size is $60 \times 60 \times 60$, and the LBM relaxation parameter is $\tau = 1$. Two walls at $z = 0$ and $z = 60$ confine the suspension. The remaining simulation parameters are $\kappa_S = 0.1$, $\kappa_\alpha = 1$, $\kappa_B = 0.01$, $\kappa_A = 1$, $\kappa_V = 1$, $\kappa_{int} = 0.05$, and $\kappa_{gl} = 0.1$. First, the system has been initialized as explained in section 8.5. The resulting system was then taken as the initial state for all four simulations mentioned above.

Velocity boundary condition

For the shear rate driven systems (rough and smooth walls), the two walls were instantaneously accelerated to a velocity of ± 0.02. The resulting wall stresses are shown in fig. 9.3(a). It can be seen that the stresses in the system with smooth walls are smaller. The reason is the low-viscosity slip layer at each wall which can maintain most of the imposed strain. As a result, the shear rate in the bulk region is smaller than the average shear rate, cf. fig. 9.4(b). This is a disadvantage because the bulk shear rate cannot be controlled *a priori* when smooth walls are used. The total stresses obtained from the MOP have been averaged between $t = 2000$ and $t = 10000$ and over

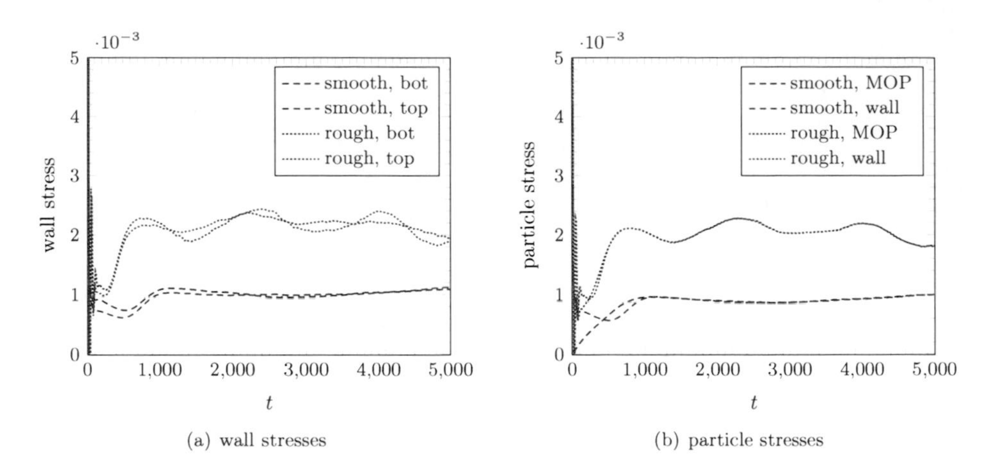

(a) wall stresses

(b) particle stresses

Fig. 9.3.: Suspension stresses for rough and smooth walls. (a) For the shear rate driven systems, the resulting wall stresses (bottom and top walls) are presented as function of time. The presence of the slip layers for the smooth walls reduces the stress: The system can be sheared more easily. (b) The particle contribution to the wall stress (averaged over bottom and top walls) is compared with the volume-averaged stress obtained from the method of planes (MOP). After the initial transient in which the fluid is accelerated, both methods yield the same results, even as function of time.

the entire volume: $\langle\sigma_{xz}\rangle_{V,t} = 2.14 \times 10^{-3}$ for the rough and $\langle\sigma_{xz}\rangle_{V,t} = 1.02 \times 10^{-3}$ for the smooth walls. The results obtained from the wall stresses (the bottom and top walls individually) are identical, even if the bottom and top wall stresses are generally not identical at a given time instance, cf. fig. 9.3(a). Batchelor's approach cannot be applied to the system with rough walls because it is not clear how to evaluate the stress related to the roughness force. Still, for the systems with smooth walls, also Batchelor's approach gives the same result for the averaged stress.

In fig. 9.3(b), it is illustrated that the volume-averaged MOP stress matches the average of bottom and top wall stresses even when plotted as function of time. The fluid stress has been subtracted from the wall stresses in order to recover the contribution of the particles. Additionally, the time evolution of the particle stress for the smooth walls obtained from Batchelor's approach matches the corresponding data obtained from the MOP. The curves are identical. Therefore, Batchelor's stress is not shown separately. The transient during the first 1000 time steps in which both stresses are not identical is caused by the required acceleration of the fluid because the fluid is initially at rest. The findings illustrated in fig. 9.3(b) impressively underline that all three methods (wall, MOP, Batchelor) recover the same (volume-averaged) stress as function of time.

Additional results obtained for the simulation with smooth walls are shown in fig. 9.5. In fig. 9.5(a), it is illustrated how important the MOP is for the correct evaluation of the *local* particle stress. If Batchelor's approach is used to compute the stress for each particle individually and then adding this contribution to the z-bin in which the particle center is located, an extremely fluctuating stress profile $\langle\sigma^{\mathrm{P}}_{xz}\rangle_{x,y,t}(z)$ is recovered. The MOP, however, yields the correct stress curve, and the total stress (sum of fluid and particle contributions) are found to be independent of the transverse coordinate z (solid line). For the same simulation, the profiles of the relative apparent viscosity and the local volume fraction are shown in fig. 9.5(b). It can be easily inferred that both Batchelor's stress profile from fig. 9.5(a) and the viscosity are correlated with the volume fraction. Due to the small system size and the short integration time, the averaged density profile shows large fluctuations. In order to extract physically meaningful results, larger systems, longer integration times, and a larger number of independent runs have to be performed,

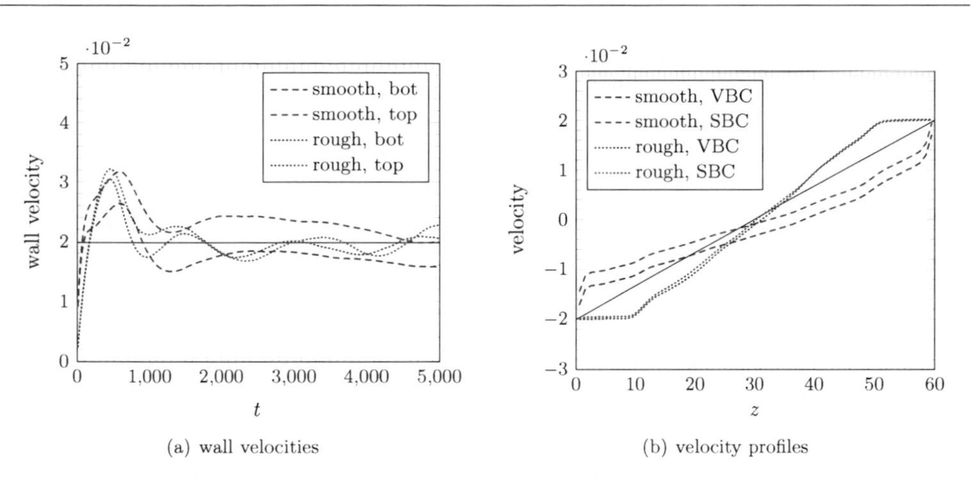

(a) wall velocities

(b) velocity profiles

Fig. 9.4.: Suspension velocities for rough and smooth walls. (a) For the shear stress driven systems, the resulting wall velocities (ignoring the minus sign for the bottom wall) are presented as function of time. The time averaged wall velocity (0.02 in all cases) is shown as solid line as reference. (b) The x-components of the velocity (averaged over the xy-plane and times between $t = 2000$ and $t = 10000$) for each simulation run (velocity boundary condition [VBC] and stress boundary condition [SBC]) are shown as function of lateral position z. The low-viscosity slip layers at the smooth walls become noticeable by the localized large velocity gradients. In contrast, for rough walls, there are extended regions (about one average particle diameter) where the velocity is basically constant. The linear velocity profile for the suspending fluid without particles is shown as solid line as reference.

cf. chap. 10. The mere intention of this section is to point out the validity and capability of the stress evaluation methods.

Shear stress boundary condition

For the shear stress driven systems, the imposed stress was chosen to be equal to the resulting stress obtained from the shear rate driven simulations in the interval between $t = 2000$ and 10000. All other simulation parameters have been the same. The resulting wall velocities are shown in fig. 9.4(a). Their averages, also taken between $t = 2000$ and $t = 10000$ give the same velocities as those which have been used for the shear rate driven simulations (± 0.02). This, again, shows the consistency of the SBC, even for rough walls.

Two observations in fig. 9.4(a) have to be explained in more detail: First, the bottom and top walls do not move with the same velocities, not even when averaged over time. The combined average velocity of bottom and top wall, however, yields the expected value of about 0.02. The reason is that, due to the small system size, spatial inhomogeneities are significant, and the viscosity is not the same close to the bottom and top walls. This is also illustrated in fig. 9.5(b). Since the shear rate and not the wall velocity is the relevant quantity, different wall velocities are not problematic as long as the shear rate is correct. Second, one can observe an overshoot of the velocity at $t \approx 500$. This can be understood from the fact that the particles have to be deformed first before they produce significant elastic stresses which decelerate the walls again. The presence of the slip layers, both for the shear rate and the shear stress driven simulations, can be easily inferred from fig. 9.4(b).

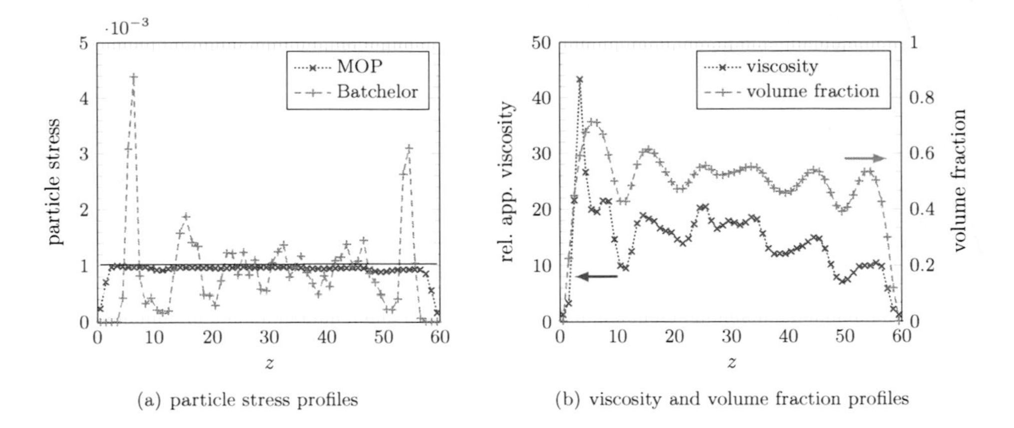

(a) particle stress profiles

(b) viscosity and volume fraction profiles

Fig. 9.5.: Suspension stress, viscosity, and volume fraction profiles. (a) The stresses (averaged between $t = 2000$ and $t = 10000$) for the shear rate driven simulation with smooth walls are shown as function of lateral coordinate z. Batchelor's approach may be used to compute the stress for individual particles and add this contribution to the bin where the particle center is located. This way, a strongly fluctuating particle stress profile is recovered which is clearly correlated to the local density of the particles as shown in (b). Instead, the method of planes (MOP) produces accurate local results for the stress, and the total stress (sum of fluid and particle contributions, solid line) is constant as expected. The results for the remaining three simulations are qualitatively similar and are not shown separately. (b) The local relative apparent viscosity and the local volume fraction for the shear rate driven simulation with smooth walls are shown as function of z. The viscosity and density peaks are obviously correlated.

Conclusions and remarks

The present simulation tool provides the possibility to drive suspensions either by shear rate (via wall velocity) or by shear stress (via wall stress). Independently of this, the walls can be made rough in order to avoid slip. This slip is caused by the presence of the liquid lubrication layers. Although the slip itself is not problematic (walls are present in either case), it is not possible to control the bulk shear rate a priori since the ratio of the viscosity of the lubrication layer and the bulk is not known in advance. The inclusion of the rough walls circumvents this problem.

The stress evaluation approaches (wall, Batchelor, MOP) yield consistent results and complement each other. It has been shown that the fluid and particle contributions to the stress can be computed locally (both in space and time). This opens the door for the analysis of spatio-temporal stress fluctuations which carry important information about the statistical properties of the system. The time-averaged stresses recover the behavior expected for a steady flow.

For all the above simulations, a zero shear stress boundary condition has been used for the y-direction (vorticity axis) at the bottom and the top wall. In particular, this means that the total momentum of the suspension in y-direction is always conserved because the momentum of the fluid can only change due to the influence of external forces like gravity (which is not the case here) or to shear stresses at the walls. Indeed, the y-component of the total fluid momentum was found to be constant up to machine precision, which is another strong indication for the proper functioning of the SBC. As a consequence the 'center of mass' of the fluid does not move along the y-axis when the initial momentum along this axis is zero. This is of paramount importance for the study of particle diffusivities (section 10.7) because undesired superimposed drift velocities may hamper the analysis otherwise.

10. Rheology and microscopic behavior of red blood cell suspensions

In this chapter, the rheology of red blood cell (RBC) suspensions is investigated numerically. In particular, the focus of the study is on the relation between the microscopic characteristics of the suspension (e.g., particle deformation, alignment, rotation, and diffusivity) and the rheology (e.g., viscosity and suspension stress). For the first time, a detailed and systematic analysis of the microscopic origins of the shear thinning behavior of blood for varying volume fractions, shear rates, and RBC deformabilities is reported.

The setup of the simulations and remarks regarding data analysis are given in section 10.1. The characterization of the dynamics of individual RBCs is introduced in section 10.2. In section 10.3, the viscosity and shear thinning behavior of the suspensions are characterized. The microscopic properties of the sheared suspensions are scrutinized in the subsequent sections: the particle rotation in section 10.4, the particle deformation in section 10.5, the collective particle alignment in section 10.6, the displacements of the RBCs in section 10.7, and the shear stress fluctuations in section 10.8.

10.1. Simulation setup and data evaluation remarks

Simulation parameters

The simulations have been performed for four hematocrit values (volume fractions), five imposed shear rates, and two particle deformabilities. In the following, all quantities are given in lattice units except indicated otherwise explicitly. The number of RBCs in the simulation box ($N_x \times N_y \times N_z = 100 \times 100 \times 160$ lattice nodes) is 494, 635, 776, and 917 for the considered volume fractions 35, 45, 55, and 65%, respectively. The applied shear rates cover two orders of magnitude between approximately 1.2×10^{-5} and 1.2×10^{-3}, resulting in inverse shear rates between about 800 and 80000. For the softer RBCs (also denoted 's' in the legends of the figures in this chapter), the parameters $\kappa_S = 0.02$ and $\kappa_B = 0.004$ have been used. For the more rigid RBCs (also denoted 'r'), the values $\kappa_S = 0.06$ and $\kappa_B = 0.012$ are taken instead. All other simulation parameters are given in tab. 10.1 and tab. 10.2. Since the ratio $\kappa_S/\kappa_B = 5$ is constant, only κ_S will be used for characterization in the following. The 2-point interpolation stencil for the immersed boundary method, eq. (6.11), and the linearized lattice Boltzmann equilibrium distributions, eq. (5.9), have been used. The employed mesh for the RBCs has 1620 faces and 812 nodes with an average distance of one lattice constant between neighboring nodes (section 8.3 and fig. 8.3(g)).

The simulation parameters for the softer RBCs have been chosen in such a way that they correspond to the physiological values of the plasma viscosity ($\eta_0 = 1.2\,\mathrm{mPa\,s}$), large RBC radius ($r = 4\,\mu\mathrm{m}$), RBC shear modulus ($\kappa_S = 5\,\mu\mathrm{N\,m^{-1}}$), and RBC bending modulus ($\kappa_B = 2 \times 10^{-19}\,\mathrm{Nm}$) [31, 234]. For this set of parameters, the capillary number

$$\mathrm{Ca} = \frac{\eta_0 \dot{\gamma} r}{\kappa_S}, \tag{10.1}$$

the numerical shear rate $\tilde{\dot{\gamma}}$ (in lattice units), and the physical shear rate $\dot{\gamma}$ (in units of s^{-1}) are

Tab. 10.1.: Parameters for simulations of red blood cell (RBC) suspensions. All parameters are given in lattice units. Values in parentheses denote deviating parameters for the more rigid RBCs.

parameter	symbol	value
system size	$N_x \times N_y \times N_z$	$100 \times 100 \times 160$
LBM relaxation parameter	τ	1
fluid density	ρ	1
large RBC radius	r	9
RBC volume modulus	κ_V	1
RBC surface modulus	κ_A	1
RBC area modulus	κ_α	1
RBC strain modulus	κ_S	0.02 (0.06)
RBC bending modulus	κ_B	0.004 (0.012)
interaction modulus	κ_int	0.05
roughness modulus	κ_gl	0.1

Tab. 10.2.: Applied wall velocities and shear rates for simulations of red blood cell suspensions. All parameters are given in lattice units. The observed shear rates in the bulk are slightly larger, cf. fig. 10.1(a).

applied wall velocity u^w	applied shear rate $\dot{\gamma}$	number of time steps $\times 10^3$	appr. number of inverse shear rates
0.00096	1.2×10^{-5}	500	6
0.00288	3.6×10^{-5}	300	11
0.0096	1.2×10^{-4}	150	18
0.0288	3.6×10^{-4}	100	36
0.096	1.2×10^{-3}	50	60

related according to

$$\mathrm{Ca} = 75\,\tilde{\dot{\gamma}}, \quad \dot{\gamma} = 78125\,\mathrm{s}^{-1}\,\tilde{\dot{\gamma}}. \tag{10.2}$$

The shear flow is wall-driven with two walls at $z = 0$ and $z = L_z$ (wall distance $L_z = 160$). The imposed boundary conditions (BCs) at the walls are velocity BCs in the x-direction and zero shear stress BCs in the y-direction (section 5.4). The wall velocities are chosen in such a way that the desired average shear rates are obtained. The remaining BCs in the x- and y-directions are periodic. In order to avoid wall slip, one layer of RBCs is glued to the walls (section 8.8). Thus, the effective bulk shear rate $\dot{\gamma}_\mathrm{eff}$ is larger than the average shear rate between the walls since the effective wall distance is decreased by about two large RBC diameters, cf. fig. 10.1. In the following, $\dot{\gamma}_\mathrm{eff}$ is abbreviated by $\dot{\gamma}$ for convenience.

Some of the simulations have become unstable and could not be evaluated. The simulations with 65% volume fraction and the wall velocity 0.096 (for the softer particles) and the wall velocities 0.096 and 0.0288 (for the more rigid particles) have been rejected.

For comparison, also a series of simulations with a single RBC have been performed. The simulation parameters are identical to those for the soft RBCs (tab. 10.1), except for the system size ($40 \times 40 \times 40$ instead of $100 \times 100 \times 160$). The wall velocities have been chosen in such a way that the capillary numbers 0.005, 0.010, 0.025, 0.050, 0.10, 0.15, 0.20, 0.25, and 0.50 are obtained. The major plane of the single cell has been initially aligned with the xy-plane of the simulation box (parallel to the walls).

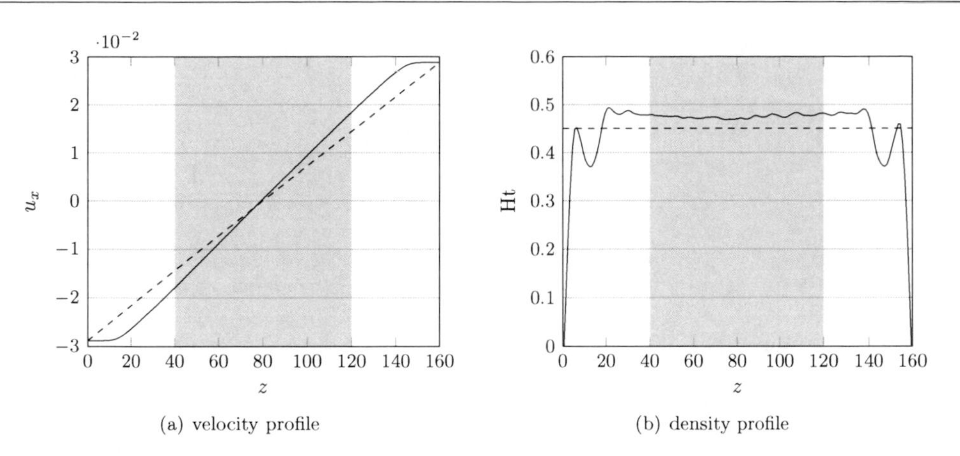

(a) velocity profile

(b) density profile

Fig. 10.1.: Velocity and density profiles for a red blood cell suspension. The solid curves are the time-averaged (a) velocity profile $u_x(z)$ and (b) density profile $\mathrm{Ht}(z)$ for the soft particles with $\mathrm{Ht} = 45\%$ and imposed shear rate of $\dot{\gamma} = 3.6 \times 10^{-4}$. The velocity gradient (shear rate) in the bulk region (between $z = 40$ and 120, denoted by gray regions in (a) and (b)) is slightly larger than the average, imposed gradient (dashed line in (a)), and the average density in the bulk region is larger than the imposed density (dashed line in (b)). The corresponding plots for the remaining simulations are qualitatively similar and are, therefore, not shown.

Ensemble averaging

For each parameter tuple $(\mathrm{Ht}, \dot{\gamma}, \kappa_{\mathrm{S}})$, five or ten independent simulation runs have been performed (ten for the softer, five for the more rigid particles). For a given tuple, the simulation parameters for each run are exactly identical, except for the initial particle position and orientation. By averaging over all runs of a tuple, the ensemble averages are improved, and statistical errors can be estimated. These errors are used to assess the statistical reliability of the results.

In practice, the observable of interest is first evaluated for each independent run alone and then averaged over all runs for the parameter tuple, if not stated otherwise. The uncertainty is defined as the root-mean-square deviation between the individual results and their arithmetic average. For a quantity Q, the individual results Q_i (which are usually already averaged over the steady state interval and bulk volume, see below) are first used to define the ensemble average based on the N independent runs,

$$\langle Q \rangle := \frac{1}{N} \sum_i Q_i. \tag{10.3}$$

The statistical uncertainty is then defined as

$$\delta Q := \sqrt{\frac{\sum_i (Q_i - Q)^2}{N}}. \tag{10.4}$$

If derived quantities such as the viscosity (as ratio of shear stress and shear rate) are reported, the uncertainty is obtained from the standard error propagation of the underlying observables.

Definition of bulk and steady state

The primary scope of the present work is the investigation of the bulk properties in the steady state. In the following, the bulk region and the steady state interval are defined.

Since the simulated volume is bounded by walls, wall effects are expected to play a role. For this reason, a bulk region is identified a posteriori before the rheological and microscopic characteristics of the suspensions are further analyzed. The bulk region is defined by the volume in which observables such as shear rate and particle density do not show significant gradients. For all simulations considered here, the region between $z = 40$ and $z = 120$ (which is the interior 50% of the total volume) is taken as the bulk, cf. fig. 10.1.

After initializing and starting a simulation, it takes a certain number of time steps until the suspension is in the steady state (in the sense that statistical properties of the system become independent of the origin of time). The duration of the transient depends on the control parameters $(\mathrm{Ht}, \dot{\gamma}, \kappa_S)$. It is not known a priori and has to be identified before further data analysis is performed. The transient may be tagged by the time behavior of observables such as the wall stress or the average particle deformation. It is noteworthy that the transient for different observables may have differing durations. For example, the transient for orientational ordering (section 10.6) is found to be longer than that for viscosity (section 10.3) or particle deformation (section 10.5). It is expected that transients are longer for observables related to collective effects as compared to individual particle properties because the time scale for structural relaxation is typically longer than for deformation of an individual particle.

Lees-Edwards boundary conditions would solve the problem of wall effects (definition of a bulk region, increased volume fraction in the bulk, etc.). However, the method of planes (section 9.4) cannot be applied when Lees-Edwards boundary conditions are used. Stress evaluation would be significantly more difficult then [235].

10.2. Characterization of particle deformation, orientation, and rotation

Fig. 10.2 shows some snapshots of the suspension configurations for various shear rates at 55% volume fraction. It can be seen that the suspension microstructure and the individual properties of the particles are different when the shear rate is changed. Motivated by this observation, it is necessary to characterize the deformation, orientation, and rotation states of the RBCs in more detail. In the following, it is explained how these quantities are defined and evaluated.

Inertia tensor

The basis for the microscopic analysis is the inertia tensor. For any extended particle, its inertia tensor $\boldsymbol{T}$ can be computed. If this particle has a constant density ρ, the volume integration in the definition of $\boldsymbol{T}$ can be easily transformed to a surface integration whose discretized version reads [187, 191]

$$T_{\alpha\beta} = \frac{\rho}{5} \sum_j^{\text{faces}} A_j \left(\boldsymbol{r}_j^2 \delta_{\alpha\beta} - r_{j\alpha} r_{j\beta} \right) r_{j\gamma} n_{j\gamma}, \tag{10.5}$$

where $\boldsymbol{r}_j$ is the vector from the particle centroid to the centroid of face j with area A_j and unit normal $\boldsymbol{n}_j$. As the tensor $\boldsymbol{T}$ is symmetric, three real eigenvalues T_i $(i = 1, 2, 3)$ can always be computed, and the diagonalized tensor reads $\boldsymbol{T} = \mathrm{diag}(T_1, T_2, T_3)$ with $T_1 \leq T_2 \leq T_3$. The inertia tensor of a particle contains valuable information: (i) Its eigenvalues allow to describe the current deformation state. (ii) The orientation of its eigenvectors characterizes the orientation of the particle in space. (iii) The change of the eigenvectors in time defines the tumbling velocity of the particle. In the following, these ideas will be elaborated on.

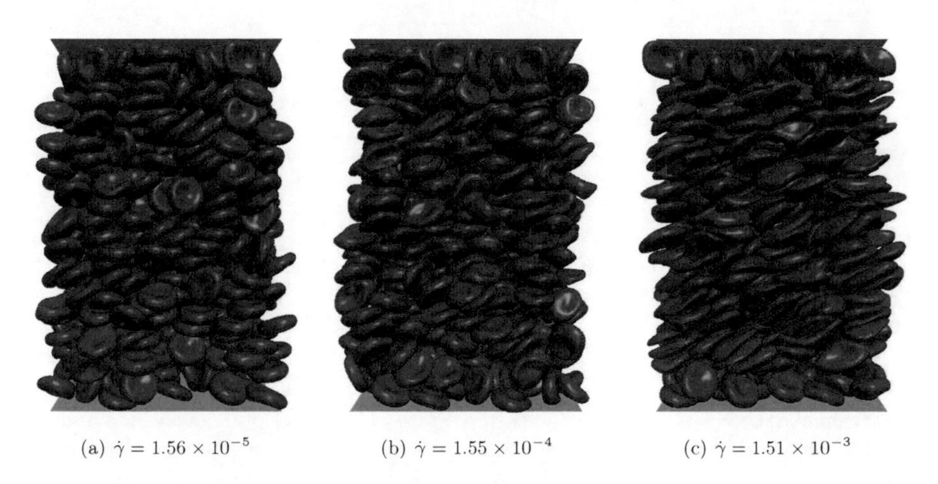

Fig. 10.2.: Snapshots of sheared red blood cell (RBC) suspensions at various shear rates. The soft RBCs at 55% volume fraction are shown in steady state for three different bulk shear rates. (a) The RBCs are more or less undeformed and behave similarly to rigid bodies. (b) Deformation becomes important. (c) The RBCs are strongly deformed.

Inertia ellipsoid

For convenience, the *inertia ellipsoid* is defined. It is the (unique) ellipsoid with the same density and inertia tensor $\boldsymbol{T}$ as the particle. Based on the inertia tensor of an ellipsoid with constant density,

$$T_1 = \frac{M(b^2 + c^2)}{5}, \quad T_2 = \frac{M(a^2 + c^2)}{5}, \quad T_3 = \frac{M(a^2 + b^2)}{5}, \tag{10.6}$$

one can show that the three semiaxes of this ellipsoid are

$$a = \sqrt{\frac{5(T_2 + T_3 - T_1)}{2M}}, \quad b = \sqrt{\frac{5(T_3 + T_1 - T_2)}{2M}}, \quad c = \sqrt{\frac{5(T_1 + T_2 - T_3)}{2M}} \tag{10.7}$$

where $M = \rho V$ is the mass and V is the volume of the particle. The semiaxes are sorted according to $a \geq b \geq c$. The inertia ellipsoid for an undeformed RBC obeys $a = b > c$. It is shown in fig. 10.3. In this particular case, $a = b = 1.1r$ and $c = 0.36r$ with r being the large radius of the RBC.

Particle orientation

Assuming a disk-like shape, i.e., a clear separation between the length of c on the one hand and the lengths of a and b on the other hand, the orientation vector $\hat{\boldsymbol{o}}$ of a particle is defined as the inertia tensor eigenvector corresponding to the shortest semiaxis c (or, equivalently, to the largest moment of inertia, T_3). Thus, the vector $\hat{\boldsymbol{o}}$ is perpendicular to the ab-plane of the particle (fig. 10.3). It has to be noted that the sign of the orientation vector is not fixed by its definition. Physical observables have to be specified in such a way that they are invariant under the transformation $\hat{\boldsymbol{o}} \to -\hat{\boldsymbol{o}}$.

Deformation parameter

A deformation parameter is introduced to monitor the deviation of the current from the equilibrium shape of a particle. A similar approach has been followed in section 8.4. In the following, the

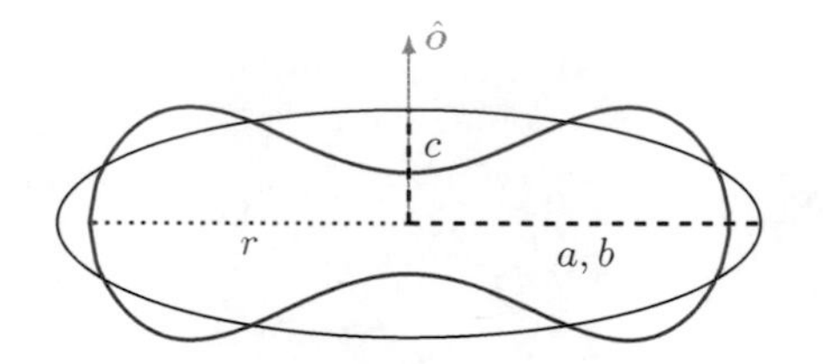

Fig. 10.3.: Red blood cell (RBC) and its inertia ellipsoid. An undeformed RBC (red) and the corresponding ellipsoid (black) with the same inertia tensor are shown to scale. The large radius of the RBC is r (dotted red line), and the principal semiaxes of the ellipsoid are $a = b = 1.1r$ and $c = 0.36r$ (dashed black lines). The orientation vector $\hat{\boldsymbol{o}}$ (gray) is perpendicular to the ab-plane.

discussion will be restricted to RBCs, or—more generally—to objects whose inertia ellipsoid has principal semiaxes $a_0 = b_0 > c_0$ in the undeformed state. The subscript 0 denotes the undeformed shape. For deformable particles, the semiaxes a, b, and c are generally time-dependent, i.e., $a = a(t)$, $b = b(t)$, and $c = c(t)$. Additionally, $a(t)$ and $b(t)$ are generally not equal, even if $a_0 = b_0$. It has to be noted that always $a(t) \geq b(t) \geq c(t)$ holds by definition.

The deviations of the current semiaxes compared to their undeformed counterparts give a first approximation of the deformation of the particle without tracking the entire surface information, which becomes unpractical if a large number of resolved particles is simulated for a long time[1]. Let $\hat{a}(t) := a(t)/a_0$, $\hat{b}(t) := b(t)/b_0$, and $\hat{c}(t) := c(t)/c_0$ be the reduced semiaxes of the deformable particle which are computed on the fly. One may then define a deformation index characterizing the asymmetry in the ab-plane (similarly to section 8.4),

$$D_{\mathrm{a}}(t) := \frac{\hat{a}(t) - \hat{b}(t)}{\hat{a}(t) + \hat{b}(t)}, \quad D_{\mathrm{a}}(t) \in [0, 1]. \tag{10.8}$$

This quantity becomes zero if the particle is undeformed.

Tank-treading and tumbling

For a rigid particle, rotational motion is always tightly connected to a rotation of its inertia tensor. Any of its mass elements rotates with the same angular velocity $\boldsymbol{\omega}$ about its center. The velocity of a mass element is found from $\boldsymbol{v} = \boldsymbol{\omega} \times \boldsymbol{r} + \boldsymbol{v}_{\mathrm{cm}}$ where $\boldsymbol{r}$ is the distance vector from the center of the particle to the particular mass element and $\boldsymbol{v}_{\mathrm{cm}}$ is the velocity of the center of mass. The angular velocity $\boldsymbol{\omega}$ at a given time is always identical for all mass elements, and it is also equal to the rotational velocity of the particle's inertia tensor, characterized by its three eigenvectors. Here, the *tumbling velocity* $\boldsymbol{\omega}$ is defined as the angular velocity of the inertia tensor in space.

Deformable particles behave differently in general. The mass elements of the particle may (i) rotate with different angular velocities, and (ii) this rotation may be independent of the rotation of the inertia tensor. A prominent example is the steady tank-treading behavior of a deformable capsule in shear flow as discussed in section 8.4. Here, the *shape* of the particle is stationary in space, i.e., the inertia tensor does not rotate, but the membrane (and with it the mass elements) rotates about this shape. A sketch of tumbling and tank-treading rotation is shown in fig. 10.4. There does not seem to be a clear definition of the instantaneous angular velocity of a deformable particle in the literature. Instead, the rotation period is commonly reported [76, 87, 236]. It is

[1]For a mesh with 1620 faces (812 nodes) and a simulation with 600 particles, the total surface information would be $3 \times 600 \times 812 \times 8\,\mathrm{B} > 11\,\mathrm{MB}$ for a single time step where a double-precision floating-point data type requires 8 bytes of computer memory. For ten independent runs and 10000 snapshots, the required data would be more than one terabyte.

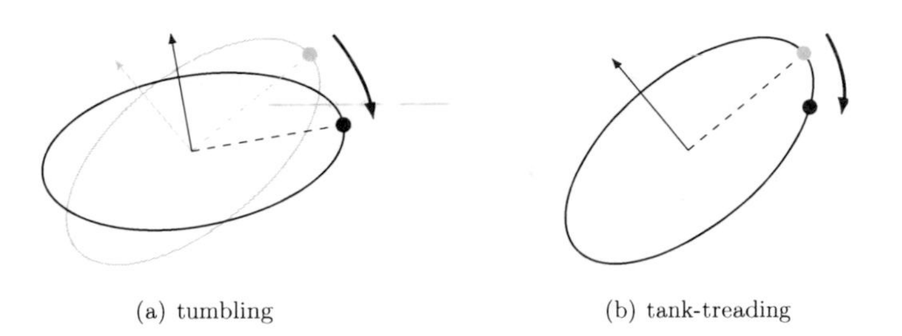

Fig. 10.4.: Tank-treading and tumbling ellipsoid. An ellipsoid is shown in (a) pure tumbling and (b) pure tank-treading state. During tank-treading, the shape of the ellipsoid does not change in space, but surface points (small circles) move along the surface. In the tumbling state, marker points do not move relatively to the ellipsoid, rather the ellipsoid itself rotates in space. The current states are shown in black, the previous states in gray, respectively. The large semiaxes (dashed) and the orientation vectors (arrows) are also shown.

found by tracking the positions of surface patches and measuring the time for one revolution. This quantity, however, is the time average over one rotation and does not provide access to the instantaneous angular velocity. Additionally, it is only reasonably well defined when the rotation is periodic, e.g., for a single particle in shear flow. For an individual particle in a dense suspension, the rotational motion may be quite erratic, and the tank-treading period is not well-defined.

It is possible to estimate the instantaneous tumbling velocity (i.e., the rotation of the shape) of particle i from the inertia tensor by tracking its eigenvector rotations in space. In the present case, it is assumed that the particle has a shape close to a disk and that the preferred rotation axis is along the y-axis (vorticity axis). The orientation vector $\hat{\boldsymbol{o}}_i$ of particle i is first projected onto the xz-plane (shearing plane) where an inclination angle $\theta_i \in [0, 2\pi[$ with respect to the x-axis can be defined. The tumbling velocity ω_i then is the time derivative of the inclination angle θ_i. This angular velocity is defined to be positive if the particle tumbles with the same vorticity as the ambient flow.

In principle, one may track the rotation of the right-handed trihedron defined by the three eigenvectors of the inertia tensor with a general rotation matrix. However, the results have been found to be imprecise in the present case. Due to the quasi degeneracy of two eigenvalues of the inertia tensor of a RBC, fast in-disk rotations of the inertia tensor can be observed even when the membrane is rotating only slowly. This is a mathematical problem which is introduced by describing the complex RBC shape only by an equivalent inertia ellipsoid. For future investigations, a more accurate approach to obtain the rotation state of the deformable particles are necessary.

Nematic ordering of disk-like particles

Whenever suspended particles are not spherical, their orientation may play a role in the rheology of the suspension. A prominent example are liquid crystals. Liquid crystals in the *nematic phase* are orientationally ordered with one preferred axis while the center positions of the particles are generally unordered [237]. The orientational order state is characterized by the *nematic order tensor* $\boldsymbol{Q}$ [237, 238, 239],

$$Q_{\alpha\beta} := \frac{1}{2}\langle 3\hat{o}_{i\alpha}\hat{o}_{i\beta} - \delta_{\alpha\beta}\rangle_{i,t}, \tag{10.9}$$

where $\hat{\boldsymbol{o}}_i$ is the orientation vector of particle i. The average is taken over an appropriate volume and time span (in the current case: bulk volume and steady state). Obviously, the signs of the

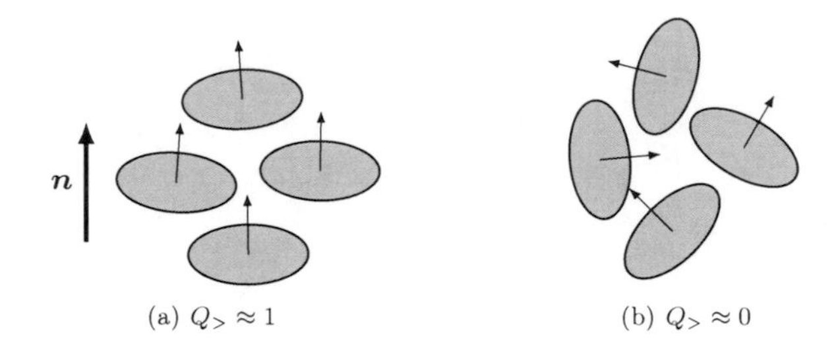

(a) $Q_> \approx 1$ (b) $Q_> \approx 0$

Fig. 10.5.: Schematics of the director $\boldsymbol{n}$ and the nematic order parameter $Q_>$. Two special cases of ordering are illustrated: (a) nearly perfect alignment and (b) nearly isotropic orientation of disk-like particles (e.g., prolate ellipsoids or red blood cells). The orientation vectors $\hat{\boldsymbol{o}}_i$ of the individual particles are shown as small arrows, the corresponding director $\boldsymbol{n}$ as thick arrow.

vectors $\hat{\boldsymbol{o}}_i$ do not play a role in eq. (10.9): The definition of the order tensor $\boldsymbol{Q}$ is invariant under transformations $\hat{\boldsymbol{o}}_i \to -\hat{\boldsymbol{o}}_i$.

The *scalar order parameter* $Q_>$ is defined as the largest eigenvalue of the order tensor $\boldsymbol{Q}$ [238]. The corresponding eigenvector is called the *director* $\boldsymbol{n}$. The director indicates the average orientation of the particles, whereas the order parameter is a measure for the amount of order: It takes the values

$$Q_> = \begin{cases} 1 & \text{if all orientation vectors } \hat{\boldsymbol{o}}_i \text{ are parallel (perfect alignment),} \\ 0 & \text{if all orientation vectors } \hat{\boldsymbol{o}}_i \text{ are randomly oriented (perfect isotropy),} \end{cases} \tag{10.10}$$

which is illustrated in fig. 10.5.

The director $\boldsymbol{n}$ and the order parameter $Q_>$ are macroscopic quantities defined in volumes containing a sufficient amount of microscopic particles. Generally, both observables are functions of position and time. In the present case, $\boldsymbol{n}$ and $Q_>$ are averaged over the bulk volume and the steady state interval since no significant dependence on position or time has been observed. This is in marked contrast to results obtained numerically for Brownian liquid crystals made of rigid oblate particles in simple shear flow where the director can be observed to rotate in space [240].

10.3. Suspension viscosity and shear thinning

The reduced apparent shear viscosity of the RBC suspensions, η/η_0, is computed from the shear stress and the bulk shear rate found during steady state via $\eta = \sigma_{xz}/\dot{\gamma}$. The shear stress σ_{xz} is obtained by fitting a constant to the profile of the suspension stress (sum of fluid and particle contributions) as obtained from the method of planes in the Eulerian frame (section 9.4). The shear rate $\dot{\gamma}$ is found by fitting a linear function to the velocity profile in the bulk region. In fig. 10.6, the viscosity η/η_0 and shear stress σ_{xz} are shown for different volume fractions and deformabilities as function of the bulk shear rate $\dot{\gamma}$. There are five main observations:

1. All curves for a given volume fraction and deformability exhibit shear thinning behavior.
2. For a given shear rate and deformability, higher volume fractions result in higher viscosities.
3. When the particle deformability is decreased (i.e., rigidity is increased), the viscosity becomes larger.
4. The Newtonian plateau at large shear rates has not yet been reached. It is expected that the viscosities will decrease further when the shear rate is increased.

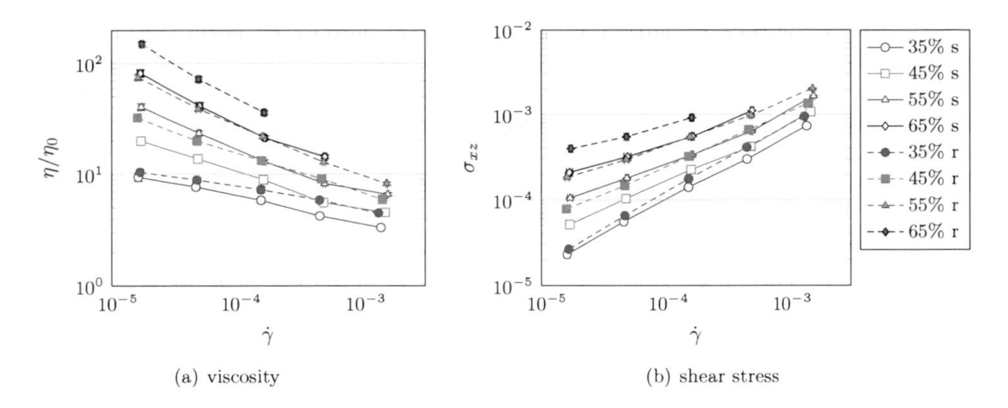

Fig. 10.6.: Viscosity of red blood cell suspensions. (a) The reduced apparent bulk viscosity η/η_0 and (b) the shear stress σ_{xz} are shown for all volume fractions and both deformabilities (softer and more rigid particles) as function of bulk shear rate $\dot{\gamma}$. Statistical errors are comparable to the symbol size. Lines are guides for the eyes.

5. The slopes of the flow curves indicate that there may be a yield stress for volume fractions $\geq 45\%$.

These points will be analyzed in the following.

Dimensional analysis and parameter reduction

The plots in fig. 10.6 reveal that the viscosity is a function of volume fraction Ht, bulk shear rate $\dot{\gamma}$, and particle deformability κ_S. It is tempting to use these three input control parameters also for the characterization of the shear thinning behavior. However, it arises the question whether these control parameters are the most appropriate ones for this purpose. Indeed, as will be shown below, two instead of three independent parameters are sufficient to describe the data.

Moreover, due to the non-linearity of the physical problem, the known *input* parameters may be not suitable to describe the *outcome* of the simulations. This well-known phenomenon is, for example, important for hard spheres: For these systems, one can define the Péclet number (ratio of the time scales for advection and bare diffusivity, e.g., diffusivity in the dilute limit) and the Weissenberg number (ratio of structural relaxation time and inverse shear rate). The former can always be defined a priori since it contains quantities which are known before the simulations or experiments are performed. The latter is only known a posteriori because the structural relaxation time strongly depends on non-linear effects. In the linear regime, the Péclet number and the Weissenberg number are proportional. However, when non-linearity is important, shear thinning is described by the Weissenberg number rather than the Péclet number [11, 19].

The first step is to identify the relevant dimensionless parameters for the present system which are based on the input parameters. Beside the volume fraction which is already dimensionless, one may define the capillary number (ratio of viscous fluid stress and a characteristic elastic membrane stress) as in eq. (10.1),

$$\mathrm{Ca} := \frac{\eta_0 \dot{\gamma} r}{\kappa_S}, \tag{10.11}$$

and the Reynolds number (ratio of inertial and viscous forces),

$$\mathrm{Re} := \frac{\rho \dot{\gamma} r^2}{\eta_0}. \tag{10.12}$$

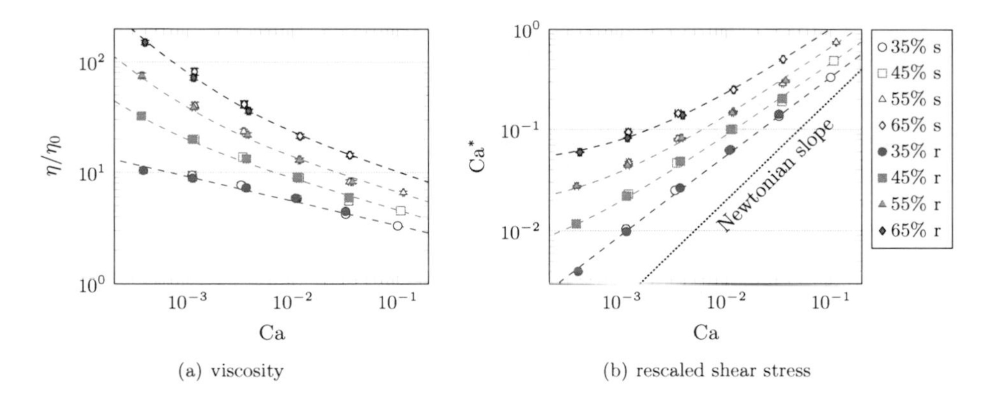

(a) viscosity

(b) rescaled shear stress

Fig. 10.7.: Rescaled viscosity of red blood cell suspensions. (a) The reduced apparent bulk viscosity η/η_0 and (b) the rescaled shear stress (in terms of capillary number $\mathrm{Ca}^* \propto \sigma_{xz}$) are shown for all volume fractions and both deformabilities (softer and more rigid particles as defined in section 10.1) as function of capillary number Ca. Statistical errors are comparable to the symbol size. The dashed lines are fits of a Herschel-Bulkley law, eq. (10.14), to the data. The fit parameters are given in tab. 10.3. A Newtonian slope ($\mathrm{Ca}^* \propto \mathrm{Ca}$, i.e., $\sigma_{xz} \propto \dot{\gamma}$) is shown as reference in (b).

In the present work, the linearized lattice Boltzmann equilibrium distributions, eq. (5.9), are employed. It is shown in appx. B.1.1 that this leads to the Navier-Stokes equations without the advective term $\rho(\boldsymbol{u} \cdot \nabla)\boldsymbol{u}$. In this sense, the Reynolds number is identically zero. However, it is also discussed in appx. B.1.1 that the explicit term $\rho\partial_t \boldsymbol{u}$ (which does not appear in the Stokes equation either) is still present. It is argued below that the formal scaling of $\rho\partial_t \boldsymbol{u}$ on the one hand and $\nabla \cdot \boldsymbol{\sigma}$ on the other hand is given by the number defined in eq. (10.12), although it cannot be interpreted as ratio of inertial and viscous forces anymore. Still, the term *Reynolds number* is kept for convenience. It is assumed that the effect of $\rho\partial_t \boldsymbol{u}$ is not important. On the one hand, for the simulations with the highest shear rate, $\mathrm{Re} \approx 0.7$, which is not necessarily negligible but clearly not larger than unity. The capillary number, on the other hand, varies between 0.0004 and 0.1. The smallness of the largest capillary number should not be misleading: Already for $\mathrm{Ca} = 0.1$, particle deformations can be significant as will be seen in sections 10.4, 10.5, and 10.6 (see also fig. 10.2(c) which corresponds to $\mathrm{Ca} = 0.1$). In other words, the shear rates chosen for the present simulations cover regions in which the particles are nearly rigid and where they are significantly deformed.

The second step is to realize that the capillary number, as defined in eq. (10.11), may not be an appropriate parameter either. Since the observed viscosities range between about $3\eta_0$ and nearly $200\eta_0$, the effective viscous stress in the fluid cannot be estimated from the reference viscosity η_0, but from the a posteriori evaluated viscosity η. Therefore, a 'corrected' capillary number is defined,

$$\mathrm{Ca}^* := \frac{\eta\dot{\gamma}r}{\kappa_S} = \frac{\sigma_{xz}r}{\kappa_S}, \tag{10.13}$$

which is the ratio of true suspension stress and a characteristic elastic membrane stress. The idea behind this definition is that the suspension stress and not the fluid stress should be responsible for the particle deformation. The particle cannot detect where the stress originates from and can only see the total stress.

In fig. 10.7, the viscosity and shear stress data is shown again, but the shear rate and the shear stress are rescaled by the deformability, $\dot{\gamma} \to \mathrm{Ca}$ and $\sigma_{xz} \to \mathrm{Ca}^*$, cf. eq. (10.11) and eq. (10.13). It can be observed that curves for different deformabilities but the same volume fraction

Tab. 10.3.: Comparison of the flow curves with a Herschel–Bulkley fluid. The viscosity data $\eta(\mathrm{Ca})/\eta_0$ and $\mathrm{Ca}^*(\mathrm{Ca})$ (fig. 10.7) can be described by the Herschel–Bulkley law, eq. (10.14). The fit parameters are given in the table below. For a Newtonian fluid, $\mathrm{Ca}_\mathrm{y}^* = 0$, $b = \eta/\eta_0$, and $p = 1$.

Ht	Ca_y^*	b	p
35%	0	1.9	0.76
45%	0.004	2.4	0.73
55%	0.015	3.4	0.72
65%	0.043	5.0	0.71

nearly collapse. This is clear evidence for the hypothesis that the viscosity depends only on two independent parameters, the volume fraction and the capillary number Ca. However, when studying properties of individual cells (e.g., rotation, fig. 10.10), Ca* turns out to be an even more relevant parameter than Ca.

It is not clear how to predict the shape of the flow curves in fig. 10.7 as function of Ht and Ca. At least, it is possible to approximate these flow curves for a given volume fraction with a Herschel-Bulkley law,

$$\mathrm{Ca}^*(\mathrm{Ca}) = \mathrm{Ca}_\mathrm{y}^* + b \times \mathrm{Ca}^p, \tag{10.14}$$

where $\mathrm{Ca}_\mathrm{y}^* = \sigma_\mathrm{y} r/\kappa_\mathrm{S}$ is proportional to the yield stress σ_y. A power $p < 1$ indicates a shear thinning fluid. The fit parameters Ca_y^*, b, and p are collected in tab. 10.3. As seen from the values of p in this table, denser suspensions are more shear thinning. It has to be noted that the Herschel-Bulkley behavior is not assumed to be valid at much larger values of Ca where a Newtonian plateau is expected. It is also risky to extract a value for the yield stress σ_y based on a limited range of shear rates. Rather, eq. (10.14) may be used as a guideline to interpret the data and to develop a theory for the investigated shear rate range. Therefore, if a yield stress really exists, its value may be different from the fit parameter σ_y.

The question arises whether a proper rescaling of the viscosity according to the volume fraction may lead to an additional collapse of the data on a single master curve as function of Ca or Ca* only. Although similar approaches for hard sphere systems exist (e.g., [1, 241]), a proper ansatz or theory for the rescaling procedure in the case of deformable particles is missing. This issue is left for future investigations.

Further remarks about the results and possible non-Stokesian effects

The curves in fig. 10.7 do not exactly collapse. Especially for large volume fractions, there is a discrepancy. There are two possible reasons:

1. For the rescaling of the particle deformability, only κ_S and κ_B have been changed by a factor of 3. The other membrane moduli (κ_α, κ_A, κ_V) and the interaction moduli (κ_int, κ_gl) have been kept constant for convenience and numerical stability reasons. Although shear and bending resistance dominate the particle shear stress, this procedure does not correspond to a perfect unit rescaling. The simulated system is slightly different from that which would have been obtained when all membrane moduli had been rescaled by a factor of 3. This is particularly the case for Ht = 65% as can be seen in fig. 10.7. Probably, the particle interaction force (section 8.7) becomes important at this volume fraction.
2. The physical length of a lattice Boltzmann time step is different for data points with the same capillary numbers but with different particle deformabilities. Artifacts due to numerical inaccuracies can therefore not be completely excluded.

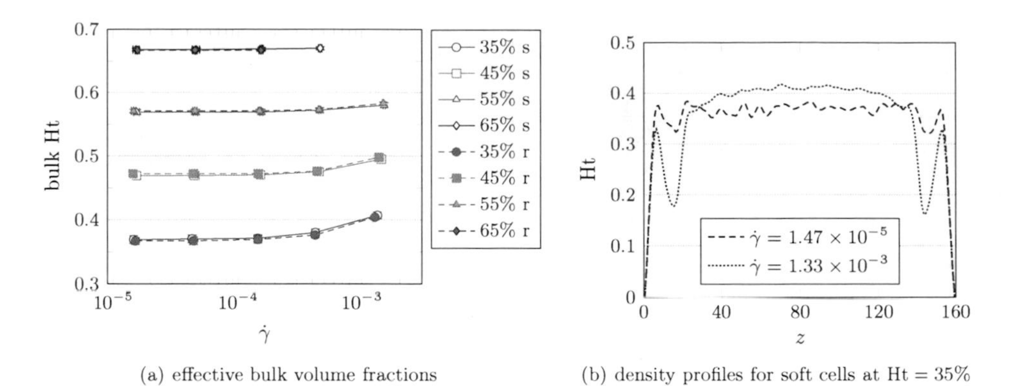

(a) effective bulk volume fractions

(b) density profiles for soft cells at Ht = 35%

Fig. 10.8.: Effective bulk volume fractions and density profiles of red blood cell suspensions. (a) The particle volume fractions (averaged over bulk volume, steady state, and independent runs) are larger than the input densities and increase with shear rate $\dot\gamma$. Error bars related to ensemble averaging are smaller than the symbols. Lines are guides for the eyes. (b) Exemplary density profiles (averaged over steady state and independent runs) are shown for the smallest and highest shear rates for the soft red blood cells with 35% input volume fraction.

In fig. 10.8(a), the time-averaged bulk densities are shown. They are larger than the target volume fractions since the density near the wall is decreased and particles are shifted towards the centerline (fig. 10.8(b)). This was to be expected. For the interpretation of the data, especially for comparisons with other simulations or experiments, the increased value of the bulk density has to be taken into account. Lees-Edwards boundary conditions would sort out this problem. However, stress evaluation as discussed in section 9.4 is strongly aggravated if not made impossible when Lees-Edwards boundary conditions are used [235].

A combination of eq. (10.12) and eq. (10.13) reveals that data points with the same value of Ca* but different deformabilities have different values of $\rho\dot\gamma r^2/\eta_0$ which is formally the Reynolds number. As mentioned before, there may be some non-Stokesian effects caused by the term $\rho\partial_t \boldsymbol{u}$. From fig. 10.8(a), it can be seen that the bulk densities increase with shear rate. This effect is stronger for smaller volume fractions. It may be related to the term $\rho\partial_t \boldsymbol{u}$ because the increase of the bulk hematocrit with shear rate always sets in at the same value of $\dot\gamma$, rather than at the same value of Ca or Ca*. One can show that a rescaling $\boldsymbol{u} \to \alpha\boldsymbol{u}$ and $\kappa_S \to \alpha\kappa_S$ leads to the same capillary numbers Ca and Ca*. However, the term $\rho\partial_t \boldsymbol{u}$ scales like α^2 because the time is also rescaled. This way, $\rho\partial_t \boldsymbol{u}$ increases faster than the remaining terms in the momentum balance equation, and it becomes more important eventually. This is a direct consequence of the fact that the LBM in its present form cannot solve the Stokes equations where the term $\rho\partial_t \boldsymbol{u}$ is absent, even when the advective term $\rho(\boldsymbol{u} \cdot \nabla)\boldsymbol{u}$ is removed by applying the linear equilibrium populations, eq. (5.9). Although the flow field is stationary on the macroscale, it is non-stationary on the microscale, and the term $\rho\partial_t \boldsymbol{u}$ cannot be neglected in the microscopic dynamics of the suspension in general. When the relative importance of $\rho\partial_t \boldsymbol{u}$ increases (especially by increasing $\dot\gamma$), hydrodynamic lift effects may arise, pushing the particles away from the wall. The significance of these lift forces is expected to be larger when the volume fraction and crowding effects are smaller, as it is observed in fig. 10.8(a). However, the study of lift forces in dense systems is not within the scope of the present thesis. As will be seen in the subsequent sections, most of the data can be accurately described by the two parameters Ht and Ca*, independently of the formal value of Re. A similar conclusion has also been drawn by MacMeccan [194] who claims that the capillary number is the only relevant parameter beside the volume fraction, even when the velocity and time time are rescaled.

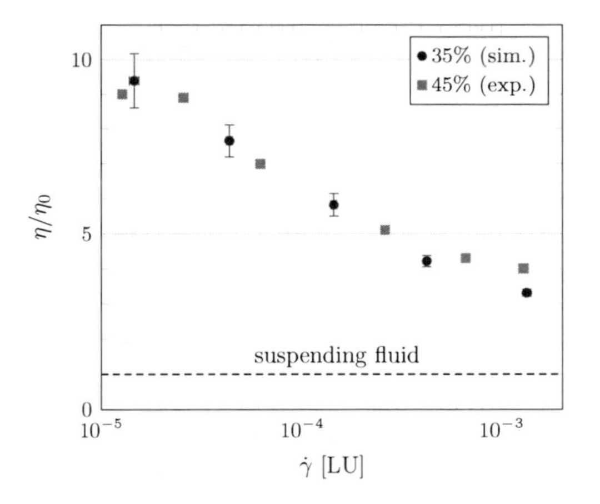

Fig. 10.9.: Comparison of blood viscosities obtained from simulations and experiments. The experimental data points are taken from [54]. They correspond to suspensions of non-aggregating red blood cells (RBCs) at Ht = 45%. The data from [54] is shown together with the simulation results for the soft RBCs at 35% volume fraction. The physical shear rates $\dot{\gamma}$ from the experiments have been converted to lattice units via eq. (10.2).

Comparison of simulation results with experiments

A comparison of the viscosity of blood obtained from the present simulations and Chien's experiments [54] is shown in fig. 10.9. It is observed that the viscosities match well over two orders of magnitude in $\dot{\gamma}$ when the simulation results for the soft RBCs at 35% volume fraction are compared with the experimentally obtained viscosities for Ht = 45%. This hematocrit discrepancy can be understood in the following way: First, the bulk volume fraction is larger than the average volume fraction, cf. fig. 10.8. Additionally, as discussed in section 8.4, the hydrodynamic radius of the particles is about $0.4\Delta x$ larger than the input radius. Therefore, the effective volume of each RBC is corrected according to $V^* = V \times C$ where

$$C \approx \frac{(r + 0.4\Delta x) \times (r + 0.4\Delta x) \times (h/2 + 0.4\Delta x)}{r \times r \times h/2} = 1.24 \tag{10.15}$$

is a volume correction factor (RBC radius $r = 9\Delta x$ and thickness $h = 6\Delta x$ for the present simulations). Therefore, the effective bulk volume fraction can be estimated by $\text{Ht}^* \approx 37\% \times 1.24 = 46\%$ which is practically the value used in the experiments. The key idea is to interpret the data in terms of an effective volume fraction which takes account of the hydrodynamic radius of the cell rather than the bare input dimension.

It can be seen from fig. 10.9 that the viscosity at larger shear rates is slightly underestimated by the simulations. The reason may be that, at these shear rates, the RBCs are tank-treading and the interior fluid contributes to the viscous dissipation. It will be shown in section 10.4 that there is indeed a transition from tumbling to tank-treading at high capillary numbers. In reality, the viscosity ratio of hemoglobin solution and blood plasma is about 5. In the simulations, it is unity. Therefore, on the one hand, the dissipation in the simulations is expected to be reduced when shear flow starts in the RBC interior. At smaller shear rates, on the other hand, the interior fluid is in a pure rotation state where no energy is dissipated, and the interior viscosity is not relevant. The viscosity of the RBC membranes (which is neglected in the present model) is also more important at higher shear stresses [203]. The inclusion of different viscosities inside and outside of the cells would be interesting for more accurate RBC simulations in the future.

According to Robertson et al. [46], the volume fraction dependence of blood viscosity decreases with increasing shear rate. Since the denser suspensions in the present simulations show a stronger shear thinning, the relative differences of the viscosities also decrease. This is nicely born out in fig. 10.7(a) where the flow curves tend to approach each other in the logarithmic representation.

10.4. Particle rotation: tumbling and tank-treading

As mentioned in section 10.2, deformable particles may exhibit tank-treading motion, i.e., rotation *without* rotating their shapes. It depends on the value of the capillary number whether the particle tumbles or tank-treads. In the limit of small Ca, the particle is virtually rigid, and it behaves similarly to a stiff object. Contrarily, when Ca is large, the particle is deformed and prefers tank-treading motion as will be argued soon.

Jeffery [242] has shown that the tumbling period for a rigid ellipsoid in simple viscous shear flow is

$$T = \left(p + \frac{1}{p}\right) \frac{2\pi}{\dot{\gamma}} \tag{10.16}$$

where $p = a/c$ (a and c are the semiaxes of the ellipsoid in the shearing plane). The value of the semiaxis b perpendicular to the shearing plane does not play a role. For a rigid sphere, $p = 1$ and the average tumbling frequency[2] is $\bar{\omega}/\dot{\gamma} = \frac{1}{2}$ where $\bar{\omega} := 2\pi/T$. The period is longer for any other aspect ratio. For the inertia ellipsoid of an undeformed RBC ($a = 1.1r$, $c = 0.36r$), Jeffery's solution predicts $\bar{\omega}/\dot{\gamma} = 0.30$. A deformable particle in pure tank-treading state has $\bar{\omega} = 0$.

It is instructive to investigate the average tumbling frequency of the RBCs in the suspension as function of the capillary number. In the following, $\bar{\omega}$ is the tumbling frequency averaged over all RBCs in the bulk during the steady state. It is expected that $\bar{\omega}/\dot{\gamma}$ should decrease when the capillary number is increased. The reason is that the particles become more and more deformable and that tumbling is replaced by tank-treading. Eventually, all particles should be in a nearly pure tank-treading state.

Numerical results and interpretation

The results for $\bar{\omega}/\dot{\gamma}$ are shown in fig. 10.10 both as function of Ca and Ca*. There are various noticeable observations:

1. All data curves collapse onto a single curve for $\text{Ca}^* > \text{Ca}^*_{\text{cr}} = 0.2$ when the data is plotted as function of Ca*. When the data is plotted as function of Ca, at least the curves for different deformabilities collapse.
2. Around Ca^*_{cr}, the tumbling frequency strongly decreases.
3. For $\text{Ca}^* < \text{Ca}^*_{\text{cr}}$, the tumbling frequency is larger for smaller volume fractions.
4. Below $\text{Ca}^* = 0.1$, the tumbling frequency *increases* with the capillary number.

The first two observations can be understood based on the discussions in section 10.3. The deformability κ_S and the bulk shear rate $\dot{\gamma}$ can be replaced by Ca, and the data can be described by two instead of three variables (Ht and Ca). This supports the hypothesis that the term $\rho \partial_t \boldsymbol{u}$ does not play a noticeable role in the present simulations. Certainly, the more interesting observation is that *only one* parameter, the corrected capillary number Ca*, is sufficient to describe all data for $\text{Ca}^* > \text{Ca}^*_{\text{cr}}$. For large capillary numbers and not too large a volume fraction, the particles are deformable enough to be in a more or less isolated tank-treading state without

[2] For a sphere, the instantaneous tumbling frequency is constant. For general ellipsoids, it is a function of time.

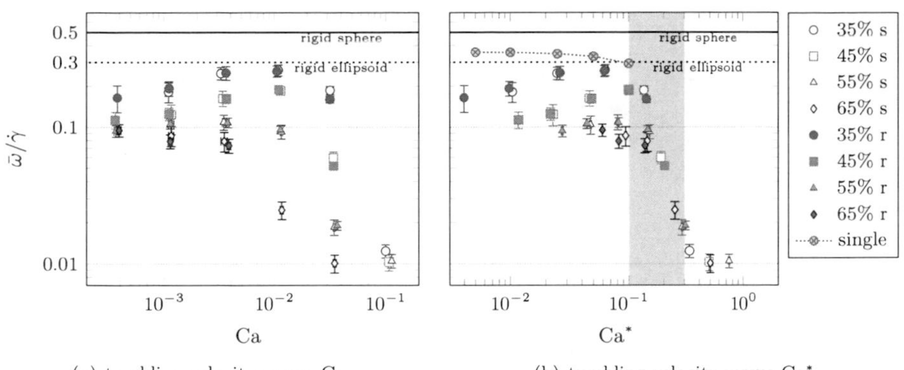

(a) tumbling velocity versus Ca

(b) tumbling velocity versus Ca*

Fig. 10.10.: Average tumbling frequencies of suspended red blood cells (RBCs). The average reduced tumbling frequency $\bar{\omega}/\dot{\gamma}$ is shown as function of capillary number (a) Ca and (b) Ca*. When plotted as function of Ca*, all data collapse on a single curve for $\text{Ca}^* > \text{Ca}^*_{\text{cr}} = 0.2$. The analytic values for a rigid sphere and a rigid ellipsoid with aspect ratio ($p = 1.1/0.36$) and the values obtained for an isolated deformable RBC are also shown. For $\text{Ca}^* \geq 0.15$, the tumbling velocity of a single RBC is zero on average. The gray area denotes the region ($\text{Ca}^* \in [0.1, 0.3]$) where tumbling is replaced by tank-treading.

the requirement to rotate their shapes additionally (thus the decay of $\bar{\omega}/\dot{\gamma}$ around Ca^*_{cr}). The particles tank-tread in their own private volume, and direct (i.e., non-hydrodynamic) collisions with neighbors are suppressed. In contrast, rigid non-spherical particles in a dense suspension necessarily have to collide during tumbling. At large Ca*, particles are aware of their neighbors only via the stresses in the fluid surrounding them. The suspension stress which is contained in the definition of Ca* seems to be the correct quantity to describe the deformation state of the particles. This hypothesis will also be supported by the results provided in sections 10.5 and 10.6. The transition from large tumbling frequencies to small values at Ca^*_{cr} marks the point at which tank-treading sets in.

The data set for the isolated RBC in fig. 10.10(b) supports this idea. For $\text{Ca}^* \geq 0.15$, a tumbling rotation of the particle cannot be detected. Instead, the particle is tank-treading and its inertia tensor does not rotate in space anymore. This explains the rather abrupt decay of the average tumbling frequency around Ca^*_{cr}. For a suspension of particles, collisions between particles lead to tumbling events for even larger values of Ca*. It is also observed that $\bar{\omega}/\dot{\gamma}$ for a nearly rigid RBC is about 20% larger than for its inertia ellipsoid ($a = 1.1r$, $c = 0.36r$). The reason may be that the cross-sections of the undeformed RBC and its inertia ellipsoid are different (fig. 10.3), which leads to a deviating rotational motion.

To this end, it is reasonable to assume that the particle rotations in the suspensions for $\text{Ca}^* > \text{Ca}^*_{\text{cr}}$ are dominated by tank-treading, only interrupted by isolated tumbling events triggered by irregularities in the ambient velocity field. Unfortunately, the simulation data is not sufficient to provide distributions of instantaneous tumbling and tank-treading velocity probabilities. It is expected that the independence of Ht may be violated when the volume fraction becomes so large that particles have to deform in order to fill the volume, even in the absence of shear.

Most probably, crowding effects are responsible for the third observation. When the volume fraction is small and the particles are still relatively rigid, $\text{Ca}^* < \text{Ca}^*_{\text{cr}}$, each particle tumbles without colliding with its neighbors. For denser systems, however, rigid body rotations are hindered by the mere presence of nearby neighbors, and crowding effects become important.

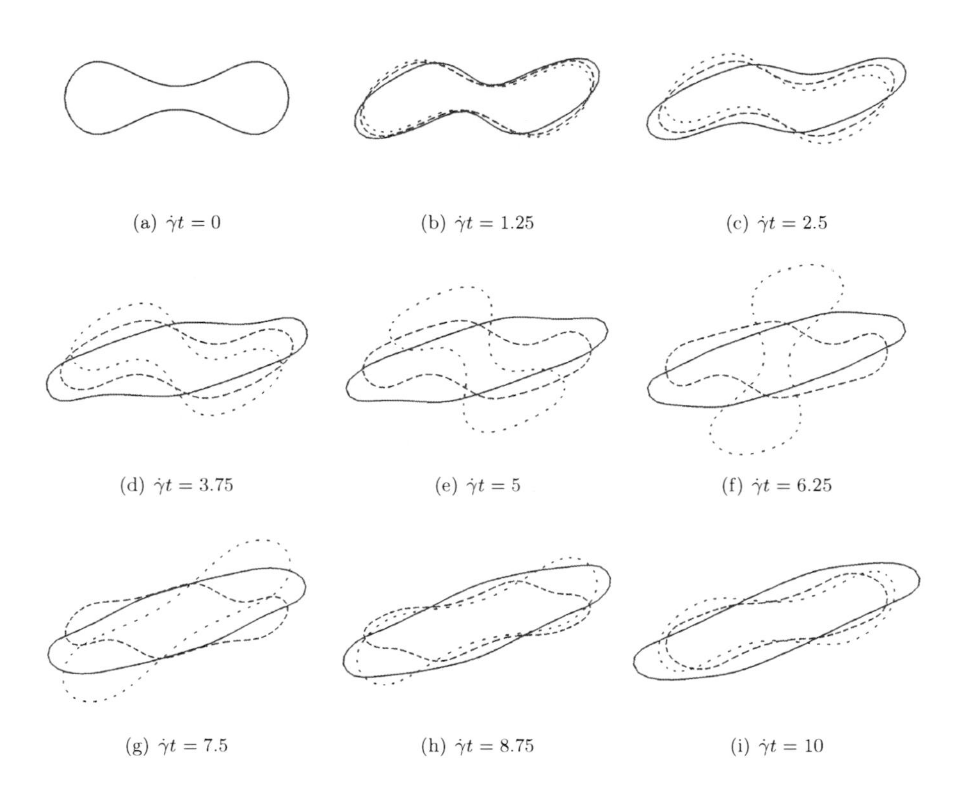

Fig. 10.11.: Tumbling and tank-treading of a single red blood cell (RBC) in shear flow. The picture sequence shows the time evolution of the rotational behavior of a single RBC in an external shear flow with shear rate $\dot{\gamma}$. The initial cell orientation at $t = 0$ is parallel to the xy-plane (perpendicular to the velocity gradient direction). The cross-section is parallel to the xz-plane (shearing plane). The vorticity of the shear flow is clockwise. The time evolution is shown for three different capillary numbers, Ca $= 0.1$ (loosely dashed line), 0.2 (densely dashed line), and 0.5 (solid line). For Ca $= 0.2$, the RBC can rotate without tumbling, whereas the RBC is not sufficiently deformed for tank-treading for Ca $= 0.1$. This becomes particularly visible at $\dot{\gamma}t = 6.25$ in (f).

Understanding the fourth observation is more difficult. One possible interpretation is that the particles are still tumbling, but with increasing Ca*, they become more deformed. When a collision between two particles during tumbling at higher Ca* occurs, the particles are, although not tank-treading, slightly softer as for smaller values of Ca* and thus squeeze past each other more efficiently. It has to be stressed again that the third and fourth observations are connected with the non-spherical shape of the RBCs.

Tumbling and tank-treading behavior of an isolated red blood cell

In order to better understand the angular velocity data in fig. 10.10, the rotational behavior of an isolated RBC has been investigated as mentioned in section 10.1. The rotation behavior of a RBC with three different capillary numbers (Ca $= 0.1$, 0.2, and 0.5) is visualized in fig. 10.11. The main observation is that the RBC for Ca $= 0.2$ is sufficiently deformable to perform tank-treading, whereas the RBC for Ca $= 0.1$ has to tumble in order to rotate. This strongly supports the interpretation that for Ca$^* \approx 0.2$, the microscopic suspension properties change

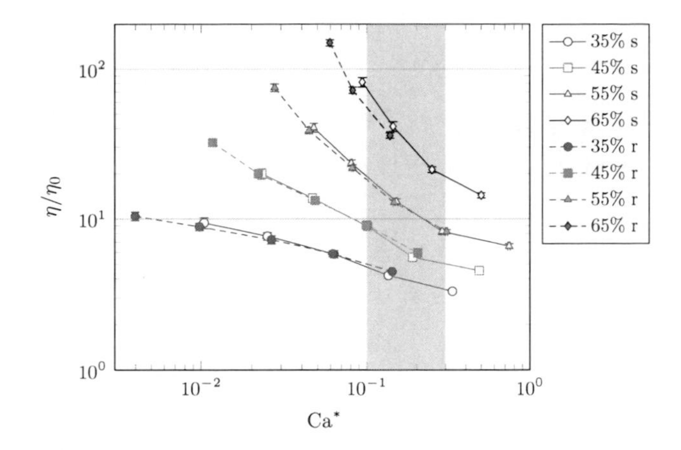

Fig. 10.12.: Rescaled viscosity of red blood cell suspensions. The reduced apparent bulk viscosity η/η_0 is shown for all volume fractions and both deformabilities (softer and more rigid particles) as function of capillary number Ca^*. The gray area denotes the region ($\mathrm{Ca}^* \in [0.1, 0.3]$) where tumbling is replaced by tank-treading.

drastically since most of the suspended RBCs are able to perform tank-treading. It should be noted that the volume fraction for an isolated RBC is small (here: $< 2\%$) and, thus, $\mathrm{Ca}^* \approx \mathrm{Ca}$.

Pozrikidis [84, 86] has found, via simulations, a qualitatively similar behavior of an isolated RBC compared to the RBC in fig. 10.11 for $\mathrm{Ca} = 0.1$. The data cannot be directly compared, though. Pozrikidis has used different constitutive membrane models, and the capillary number is defined in a slightly different way. A numerical analysis of the tank-treading behavior of isolated RBCs in simple shear flow in the limit of large capillary numbers ($\mathrm{Ca} > 0.5$) is provided by Sui et al. [243].

Effect of tank-treading on suspension viscosity

The quasi absence of tumbling-induced direct collisions of the RBCs at $\mathrm{Ca}^* > \mathrm{Ca}^*_{\mathrm{cr}}$ should be one contributing factor for the shear thinning behavior of the suspensions. Fig. 10.12 reveals that the shear thinning between $\mathrm{Ca}^* = 0.1$ and 0.3 is significant for all volume fractions. In this interval, the relative apparent viscosities decrease by a factor of about 1.7, 1.8, 2.5, and 4.0, for $\mathrm{Ht} = 35\%$, 45%, 55%, and 65%, respectively. Contrarily, the shear thinning at smaller capillary numbers should be related to other effects since tank-treading is not important below $\mathrm{Ca}^* = 0.1$. One possible mechanism for shear thinning before tank-treading becomes important may be the slight deformation of the particles as indicated before. Particles, although still tumbling, may squeeze past each other more easily when they are more deformable. In this case, another Newtonian regime at even smaller Ca^* should be observed where the particles are virtually rigid. However, as already mentioned in section 10.3, there may be a finite yield stress which could also be augmented by the presence of the repulsion force between the RBCs (section 8.7). Additional (and expensive) simulations at even smaller shear rates are required to distinguish between these two effects.

10.5. Particle deformation

In fig. 10.13, some exemplary deformation probability distributions $p(D_\mathrm{a})$ are shown. $p(D_\mathrm{a})\delta D_\mathrm{a}$ is the probability of finding a particle with a deformation as defined in eq. (10.8) in the interval

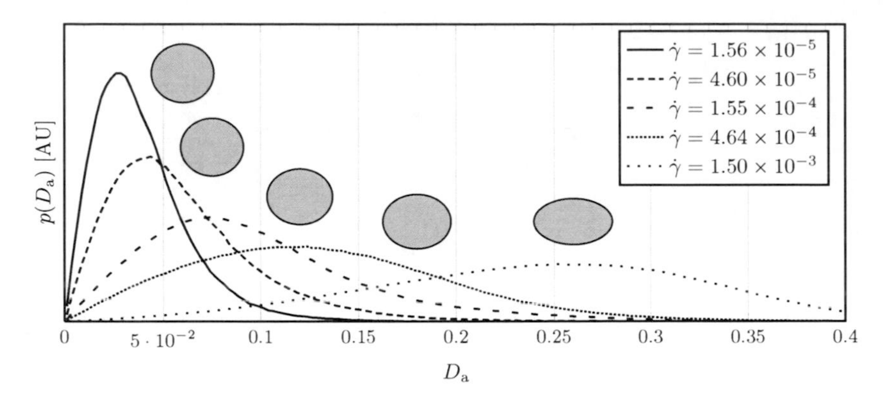

Fig. 10.13.: Deformation probability distributions for suspended red blood cells (RBCs). The deformation probability distribution for soft RBCs at Ht = 55% is shown for five different shear rates $\dot{\gamma}$. The ellipsoids denote the corresponding deformations D_a^{max} with the largest probability, respectively.

$[D_a, D_a + \delta D_a]$ in the steady state and the bulk volume. The shape of this probability distribution function generally depends on the three control parameters $(\mathrm{Ht}, \dot{\gamma}, \kappa_S)$. The major property of each distribution $p(D_a)$ is its maximum at D_a^{max}. This maximum denotes the most probable deformation of a RBC in the suspension. It is found by fitting a Gaussian to the curve in the vicinity of the maximum.

Fig. 10.14 collects all values of D_a^{max} as function of both Ca and Ca*. It is found that curves for the same volume fraction collapse when plotted as function of Ca. This, once again, supports the idea that the term $\rho \partial_t \boldsymbol{u}$ is not relevant. More striking is the observation that also curves for different volume fractions nearly collapse when plotted as function of Ca*. Obviously, Ca* is the more suitable parameter, also for the rotational behavior in section 10.4. The interpretation is that the deformation of a particle is dominated by the ambient suspension stress. The effect of the volume fraction is already contained in the capillary number Ca* through the viscosity η.

The data points also collapse for $\mathrm{Ca}^* < \mathrm{Ca}^*_{cr}$, i.e., the transition from tumbling to tank-treading described in section 10.4 is not visible in fig. 10.14. Obviously, the transition does not involve a significant change of the deformation parameter. From fig. 10.14, it can be inferred that the most probable deformation parameter at Ca^*_{cr} is $D_a^{max} \approx 0.1$ corresponding to an aspect ratio $a/b \approx 1.2$.

Interestingly, a simple scaling law quantitatively describes all deformation states $D_a^{max}(\mathrm{Ca}^*)$. Here, it is worth to take the ratio a/b of the in-plane semiaxes instead of the deformation parameter D_a. The reason is that a/b is not bounded above and can be described by a simple power law,

$$\left.\frac{a}{b}\right|_{max} - 1 = 0.89\,\mathrm{Ca}^{*\,0.90}, \tag{10.17}$$

where $a/b|_{max}$ corresponds to D_a^{max}. The fit is also shown in fig. 10.14. It should be noted that a/b can be converted to D_a and vice versa according to

$$D_a = \frac{a/b - 1}{a/b + 1}, \quad \frac{a}{b} = \frac{1 + D_a}{1 - D_a}, \tag{10.18}$$

cf. eq. (10.8). The interpretation of eq. (10.17) is that, for $\mathrm{Ca}^* = 0$, there is no deformation ($a/b = 1$ and $D_a = 0$). This relation can only be valid below a critical volume fraction above which particles have to deform to fill the volume even in the absence of shear flow. When Ca* is increased, particles are deformed in such a way that a/b grows. It is not known why the most

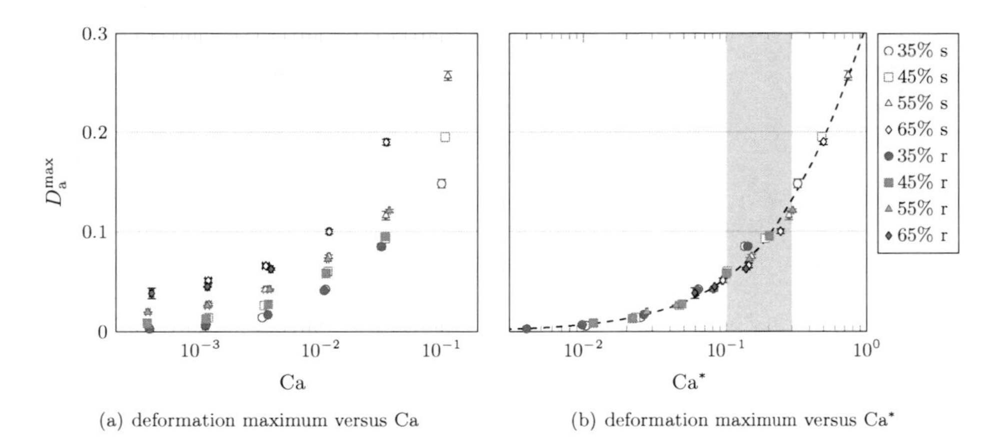

(a) deformation maximum versus Ca

(b) deformation maximum versus Ca*

Fig. 10.14.: Deformation maximum of suspended red blood cells. The most probable deformations, D_a^{max}, are shown as function of capillary number (a) Ca and (b) Ca*. The curves for different volume fractions nearly collapse on a single master curve when plotted against Ca*. The power-law fit from eq. (10.17) is shown in (b) as dashed line. The gray area denotes the region ($\mathrm{Ca}^* \in [0.1, 0.3]$) where tumbling is replaced by tank-treading.

probable deformation obeys the power-law. For future work, it may be rewarding to analyze the entire probability distributions (fig. 10.13) as function of Ht and Ca*. It also seems that the single parameter D_a is too simplistic for the description of a RBC and its various deformed shapes. Thus, a more elaborate approach for the RBC deformation should be considered in the future.

10.6. Particle alignment and orientational ordering

The director inclination angle θ and the order parameter $Q_>$ averaged over the bulk volume and the steady state are shown in fig. 10.15 as function of both Ca and Ca*. The inclination angle θ is defined as the angle between the director $\boldsymbol{n}$ (which is chosen to point into positive z-direction, i.e., $n_z > 0$) and the x-axis. The y-component of the director is basically zero at all times, $n_y \approx 0$ (data not shown). Therefore, $\tan\theta \approx n_z/n_x$. Also the x- and z-components of the director do not significantly fluctuate in time. The curves obtained for the isolated RBC in shear flow are also shown as comparison. As already seen in sections 10.4 and 10.5, the major finding is that Ca is not the appropriate parameter to interpret the data. Rather, the corrected capillary number Ca* is the relevant quantity. When plotted against Ca*, curves for different deformabilities and, depending on the value of Ca*, also curves for different Ht collapse. The results for the order parameter $Q_>$ and the inclination angle θ are described and analyzed below.

The distributions of particle inclination angles for the soft RBCs at 55% volume fraction is shown in fig. 10.16. As comparison, the probability of finding a rigid ellipsoid (aspect ratio $p = a/c$) with inclination angle θ in the shear flow,

$$p(\theta) \propto \frac{\mathrm{d}t}{\mathrm{d}\theta}(\theta) \propto \frac{p + \frac{1}{p}}{p\cos^2\theta + \frac{1}{p}\sin^2\theta}, \tag{10.19}$$

is also shown for the special case $a = 1.1r$ and $c = 0.36r$. Only for a spherical shape ($p = 1$), the probability is independent of the inclination angle. Eq. (10.19) can be inferred from Jeffery's

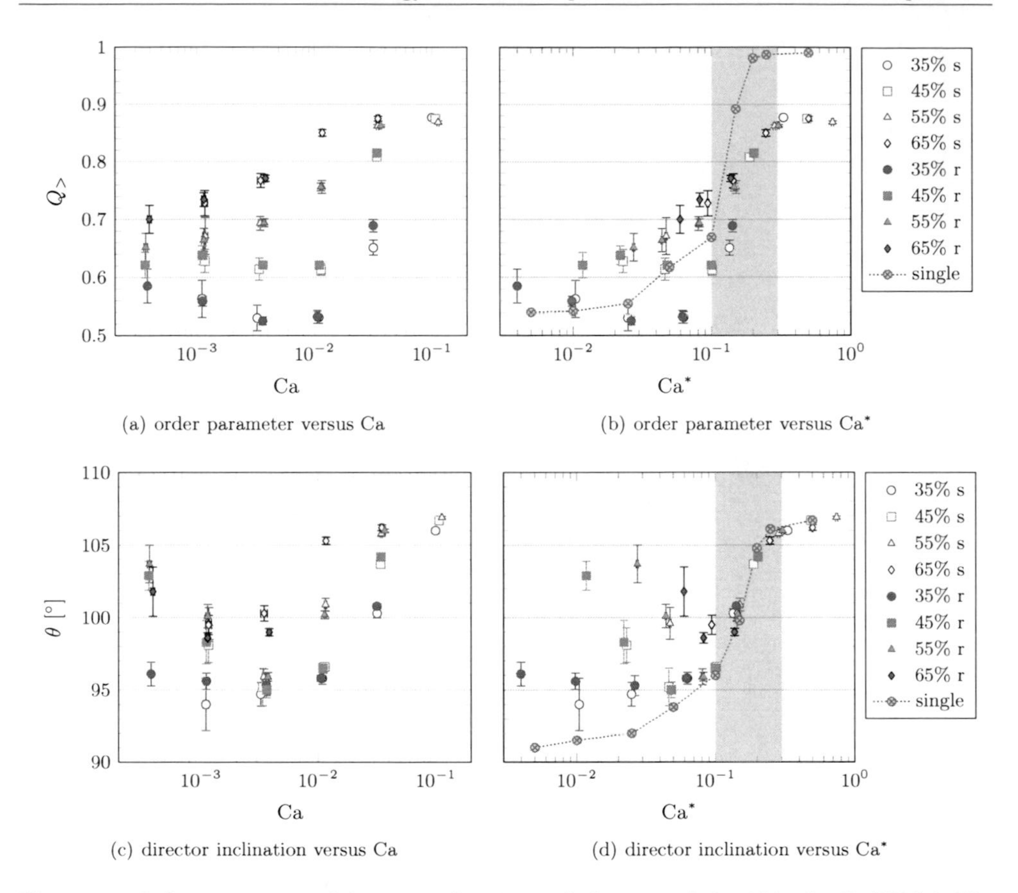

(a) order parameter versus Ca

(b) order parameter versus Ca*

(c) director inclination versus Ca

(d) director inclination versus Ca*

Fig. 10.15.: Order parameter and director inclination angle for suspended red blood cells (RBCs). The inclination angle θ of the director and the corresponding order parameter $Q_>$ are shown as function of both Ca and Ca*. The curves obtained for one single RBC are also shown (lines are guides for the eyes). The gray area denotes the region ($\mathrm{Ca}^\star \in [0.1, 0.3]$) where tumbling is replaced by tank-treading.

solution for the time evolution of the inclination angle [242],

$$\tan\theta = p\tan\left(\frac{\dot{\gamma}t}{p+\frac{1}{p}}\right). \tag{10.20}$$

Discussion of the order parameter

The behavior of the order parameter $Q_>$ in fig. 10.15(a) and fig. 10.15(b) can be summarized by four main observations:

1. For $\mathrm{Ca}^\star > 0.1$, $Q_>$ steadily increases until it reaches a plateau.
2. For $\mathrm{Ca}^\star < 0.05$ and $\mathrm{Ht} = 35\%$, $Q_>$ slightly decreases.
3. Denser suspensions show a stronger ordering for small Ca*.
4. For $\mathrm{Ca}^\star > \mathrm{Ca}^\star_{\mathrm{cr}} = 0.2$, $Q_>$ becomes independent of volume fraction.

The first observation may be interpreted in the following way: For small Ca*, the particles are basically rigid. Due to the externally imposed shear flow, particles have to rotate eventually.

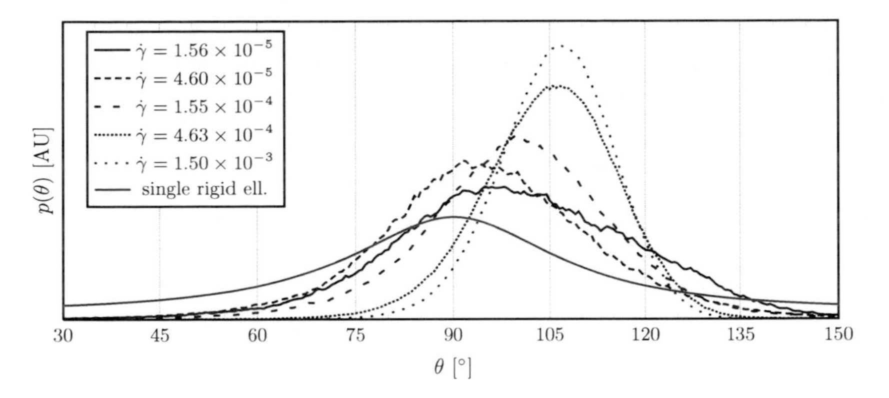

Fig. 10.16.: Inclination angle probability distributions for suspended red blood cells (RBCs). The curves denote the probability density of finding a soft RBC at 55% volume fraction with a given inclination angle in the shearing plane (xz-plane) with respect to the x-axis for five different shear rates. The director inclination angle corresponds to the position of the maximum of the curves. The analytic curve, cf. eq. (10.19), for a single rigid ellipsoid (aspect ratio $p = 1.1/0.36$) is also shown.

During this tumbling rotation, the orientational order is decreased since the orientation $\hat{\boldsymbol{o}}$ of a given number of particles is not conform to the director $\boldsymbol{n}$. For increasing Ca*, particles become softer and start to perform tank-treading. During tank-treading, particle membranes rotate without the need for tumbling, i.e., $\hat{\boldsymbol{o}}$ for a given particle does not significantly change in time. Therefore, the particles violate the orientational order less, and the order parameter increases with Ca*. At some point (at Ca$^* \approx 0.3$), a plateau with $Q_> \approx 0.88$ is reached, and the order parameter does not increase further. One reason is that even a single RBC cannot tank-tread with a spatially fixed direction $\hat{\boldsymbol{o}}$ because its dimples break the spherical symmetry of the membrane. This can be seen in fig. 10.11 and fig. 10.15(b). Additionally, fluctuating stresses in the complex flow field act on the membranes and slightly shake the orientation vectors. The second distortion mechanism seems to be more important because the curve for a single RBC approaches $Q_> \approx 1$.

The second observation may be understood in a similar way: For small but increasing Ca*, the particles are still essentially rigid, yet small deformations are more and more possible. However, tank-treading is still not relevant. The small but increasing tendency to deform may promote tumbling because colliding particles may squeeze past each other instead of becoming stuck. As a result, the ordering decreases slightly.

When the particles are still rigid (i.e., for small Ca*), they have to tumble, which is prevented by the close proximity of neighbors. On the one hand, when the suspension is more dilute, particles may rotate freely without colliding with their neighbors. This leads to a reduction of the order parameter since less particles are aligned with the director at a given time. On the other hand, for denser systems, the presence of neighbors disturbs the particles' ability to rotate freely, and the order parameter is increased. This qualitatively explains the third observation.

The fourth observation is that there seems to be a transition at $\text{Ca}^*_{\text{cr}} \approx 0.2$ beyond which all data points collapse on a single curve, although they belong to different volume fractions. One may interpret this finding by assuming that particles do not see each other anymore, except for effects which are taken into account via Ca*. Each tank-treading particle has its own volume for rotation, and there is no need for direct (non-hydrodynamic) collisions with the neighbors. For larger volume fractions, the mere effect of the dense packing is an increase of the suspension stress σ the particles feel locally. Additional evidence for this assumption has already been discussed in sections 10.4 and 10.5. Indeed, the data for the single RBC indicates that the particle can be considered rigid for $\mathrm{Ca}^* < 0.025$ and tank-treading for $\mathrm{Ca}^* > 0.2$. In the transitional interval, $0.025 < \mathrm{Ca}^* < 0.2$, the particle dynamics is more complicated since neither tumbling nor tank-treading are dominating.

Discussion of the inclination angle

The behavior of the director inclination angle θ as function of capillary number and volume fraction as shown in fig. 10.15(c) and fig. 10.15(d) can be summarized as follows:

1. For small Ca^*, θ is decreasing.
2. The inclination angle is larger for denser systems as long as Ca^* is small.
3. For larger Ca^*, the inclination angle increases again.
4. For $\mathrm{Ca}^* > \mathrm{Ca}^*_{\mathrm{cr}} = 0.2$, θ becomes independent of volume fraction.

The physical meaning of the director inclination angle θ is less obvious than that of the order parameter $Q_>$. There does not seem to be a theory for the behavior of the inclination angle for deformable RBCs at varying capillary numbers. At this point, no explanation for any of the first two observations can be given.

Comparing the data for the suspensions with the single particle curve reveals that all data collapses for $\mathrm{Ca}^* > 0.1$. Therefore, the particles in the suspension seem to behave as isolated particles, although the order parameter is smaller (fig. 10.15(b)). It should be noted that the limiting value for the inclination at vanishing capillary number is $\theta = 90°$.

The last observation can be interpreted in the same way as the similar observation for the order parameter. When the capillary numbers are large, the particles perform tank-treading and do not see their neighbors except for effects completely contained in the suspension stress and thus Ca^*. Additional direct effects due to crowding seem to be absent. This raises the question up to which volume fraction this behavior can still be observed. At some larger value for Ht, particles have to touch even when they are tank-treading.

Liquid crystals and red blood cell suspensions

Due to its significant orientational ordering, the RBC suspensions investigated in this work can formally be considered a liquid crystal. However, there are pronounced differences between a 'classical' liquid crystal and the present system. In the former case, the particles are macromolecules and orientational ordering can be observed in the absence of shear. It is a function of molecule shape, volume fraction, and temperature. In the present case, the particles are strongly deformable and orders of magnitude larger than molecules, and the thermal Péclet number is infinite. All effects, including orientational ordering, are shear induced. Therefore, the present RBC suspensions cannot be directly compared to classical liquid crystals.

The rheology of liquid crystals is generally not well-understood [8]. Although Newtonian properties are assumed in some theories [244], shear thinning behavior of liquid crystals has been observed experimentally [245, 246, 247] and in simulations [240].

Comments on the alignment and ordering of deformable and rigid particles

The reason for the minimum of the inclination angle θ at intermediate values for Ca^* (fig. 10.15(d)) may be related to the deformability of the cells and the definition of the director: At small Ca^*, particles are undeformed, and the orientation vector $\hat{\boldsymbol{o}}$ is uniquely defined. At high Ca^*, particles are basically elongated ellipsoids, and the orientation vector again is well defined. In between, the particle deformation can be more erratic, making a simple orientation vector definition more complicated. This becomes also clear from fig. 10.11 where instantaneous shapes of RBCs in shear flow are shown. It has to be remembered that, for deformed RBCs, there may be different definitions of the orientation vector. Therefore, a quantitative discussion of the alignment and ordering of RBCs in shear flow may depend on these definitions. As mentioned in section 10.4, a more accurate description of the RBC deformation (and, thus, orientation) may provide additional information in the future.

It may also be possible that the inclination angles for small Ca^* and volume fractions $\geq 45\%$ (fig. 10.15) do not reflect the steady state. As will be discussed in section 10.8, there is evidence that the transient time for the particle alignment is larger than the simulation time for these simulations. Therefore, it may be possible that longer simulation runs reveal a different inclination angle behavior for small capillary numbers.

Janoschek et al. [248] have used a simplified model for blood flow. Although individual particles are resolved, these are rigid discoid ellipsoids with the ability to overlap to some extend. This overlap is intended to mimic the deformability of the particles. However, tank-treading is not considered. The particles can only perform tumbling rotations. Although shear thinning behavior is recovered, the individual and collective particle dynamics differ from the present results. Janoschek et al. [248] observe that both the director inclination angle and the order parameter decrease with increasing shear rate. This is in marked contrast to the findings in this section. It is not surprising that models without intrinsic tank-treading ability show a different shear rate-dependence for the order parameter. Shear thinning, on the one hand, is a relatively general property of dense suspensions, irrespective of their microscopic constitution [3]. If, on the other hand, also the microscopic behavior of the constituents shall be reproduced, tank-treading seems to be unavoidable for blood simulations at higher shear rates.

10.7. Particle displacements: 'ballistic' and diffusive motion

The statistical RBC motion may be described by the mean square displacement (MSD),

$$\mathrm{MSD}_\alpha(\Delta t) := \left\langle (C_{i\alpha}(t + \Delta t) - C_{i\alpha}(t))^2 \right\rangle_{\mathrm{run},i,t}. \tag{10.21}$$

The average is taken over all independent runs, all particles i (which are in the bulk at time t) and time (in such a way that t and $t + \Delta t$ are in the steady state interval). $C_{i\alpha}$ is the α-component of the centroid of particle i. The MSD indicates which average squared distance a particle has moved along direction α in the time interval Δt. For an unsheared Brownian hard sphere system in the absence of a suspending fluid, the MSD is quadratic in Δt for small Δt and linear for large Δt. The former regime is called *ballistic* where the particles move with constant velocities between collisions. The latter is called *diffusive* and characterized by a thermal diffusion parameter D_{th}. As long as the particles are spherical and the system is homogeneous and not sheared, diffusion far away from any wall is isotropic.

For sheared systems, the diffusion mechanism depends on the shear rate. As long as the Weissenberg number is small, the system is in the linear response regime. For increasing Weissenberg numbers, the Brownian contributions decrease, and diffusion becomes mainly shear-induced. If the system is non-Brownian, as in the present case, shearing is the only mechanism for diffusion [1]. The concept of shear-induced diffusion and its experimental measurements are thoroughly described by Breedveld [94]. Diffusion in colloidal systems is reviewed in [13]. Shear-induced diffusion was first investigated by Eckstein et al. [249] and later by Leighton and Acrivos [250]: Particles move from regions with large to those with small stresses, trying to restore equilibrium which is distorted by the shear flow [13]. Similarly to thermal diffusion, the MSD is known to grow linearly for the long-time shear-induced motion [251]. Due to the presence of neighbors, particles cannot move on straight lines when the suspension is sheared, and non-affine displacements are observed which eventually give rise to the diffusive motion.

The present suspensions are different from, e.g., molecular dynamics simulations of hard sphere systems in many aspects, and it arises the question if the MSD can provide useful information for the RBC suspensions as well. On the one hand, the particles are deformable and not spherical. This renders the definition of $\boldsymbol{C}_i$ not unique (see below). On the other hand, particles in the present system interact constantly via hydrodynamic stresses whereas hard spheres in the absence of a suspending fluid do not interact between collisions and move with constant velocity. This is the reason for the term 'ballistic motion'. Due to the existence of the smooth particle interaction force (section 8.7), the presence of the dissipating suspending fluid, and the possible deformation of particles during contact, it cannot be expected that the MSD provides information which can be directly compared to results obtained from hard sphere simulations. Anyway, Bishop et al. [252] claim that shear-induced diffusion in blood vessels may increase the radial dispersion of particles and solutes by orders of magnitude as compared to Brownian diffusion.

According to Breedveld et al. [253], the shear-induced diffusion is anisotropic, i.e., a symmetric diffusivity tensor $\boldsymbol{D}$ is introduced. All of its components obey $D_{\alpha\beta} \propto \dot{\gamma} r^2$. In the remainder of this section, only the diffusivities along the y-axis (vorticity direction), D_{yy}, and the z-axis (velocity gradient direction), D_{zz}, are considered. They are abbreviated by D_y and D_z, respectively.

Computation of the mean square displacements

The first step to obtain the MSD is the definition of the centroid of particle i. Two definitions have been tested in the present work. The first is the center of the surface,

$$\boldsymbol{C}_i^{\mathrm{A}} := \frac{\int_{A_i} \mathrm{d}A\, \boldsymbol{x}}{\int_{A_i} \mathrm{d}A}, \tag{10.22}$$

the second is the center of the volume (center of mass for constant density),

$$\boldsymbol{C}_i^{\mathrm{V}} := \frac{\int_{V_i} \mathrm{d}V\, \boldsymbol{x}}{\int_{V_i} \mathrm{d}V}, \tag{10.23}$$

where $\boldsymbol{x}$ is a point either on the surface or in the volume of particle i, respectively. For spheres and other symmetric objects (such as an undeformed RBC), both definitions are equivalent. However, for an asymmetric particle (e.g., a deformed RBC), both centroids may be located at different points in space. It is not directly obvious which definition is more reasonable because the mass of the RBCs does not play a role. Therefore, both definitions have been used separately.

It turns out that the definition of the MSD as given in eq. (10.21) has to be corrected for finite size effects in order to produce more reliable results. Due to the finite system size, a non-negligible linear particle displacement along the vorticity direction can generally be observed. Therefore, the MSD is computed from

$$\mathrm{MSD}_\alpha(\Delta t) = \left\langle \left(C_{i\alpha}(t+\Delta t) - C_{i\alpha}(t) - \mathrm{LD}_\alpha(\Delta t)\right)^2 \right\rangle_{\mathrm{run},i,t} \tag{10.24}$$

where

$$\mathrm{LD}_\alpha(\Delta t) := \langle C_{i\alpha}(t+\Delta t) - C_{i\alpha}(t) \rangle_{\mathrm{run},i,t} \tag{10.25}$$

is the average linear displacement of the particles within the time interval Δt. The reason for the linear displacement can be understood in the following way: The total momentum and with it the average velocity of the fluid in the vorticity direction is conserved because (i) the NSE

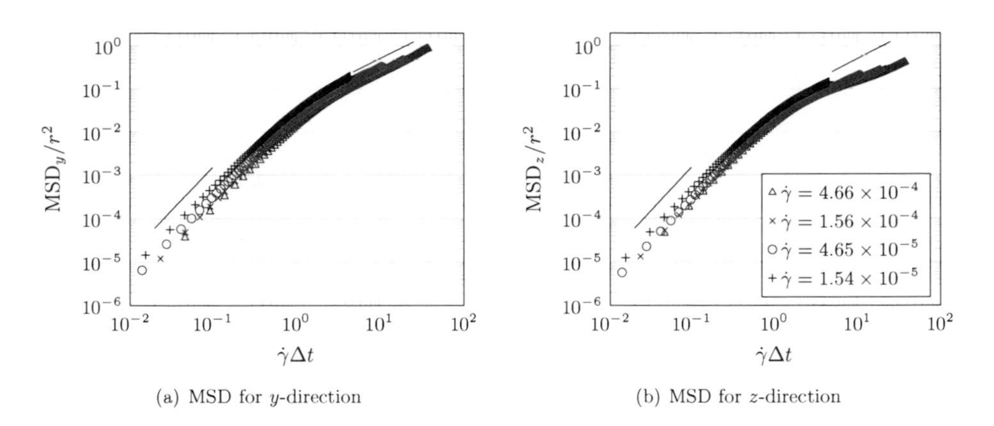

(a) MSD for y-direction

(b) MSD for z-direction

Fig. 10.17.: Mean square displacements (MSDs) of suspended red blood cells (RBCs). The MSDs for the soft RBCs at 65% volume fraction based on the volume centroid are shown for different applied shear rates $\dot{\gamma}$ and for the motion along the (a) vorticity and (b) velocity gradient direction. The MSDs have been normalized by the large RBC radius $r = 9$, the time axis by the inverse shear rate $1/\dot{\gamma}$. Curves for the other volume fractions and the more rigid RBCs are qualitatively similar and are not presented. The solid lines denote the asymptotic Δt^2- and Δt-slopes, respectively.

and the LBM are momentum conserving and (ii) a no stress boundary condition at the wall in vorticity direction (section 5.4.2) has been chosen. However, a net momentum exchange between the bulk and the wall regions generally takes place, leading to a generally non-zero velocity of the bulk region. This average motion may mask the fluctuating motion of the particles if not filtered accordingly. For the motion in velocity gradient direction (along the z-axis), the linear displacements are less relevant because motion is strongly hindered by the presence of the walls at $z = 0$ and $z = L_z$.

Investigations of the present data have shown that the results for the MSD are only acceptable when the averaging volume is as large as possible, even when the corrected version in eq. (10.24) is used. Therefore, the data is not sufficient to allow the study of the z-dependence of the MSD by dividing the volume between the walls into smaller bins. In particular, the near-wall behavior of the MSD cannot be analyzed as compared to the bulk behavior. The number of RBCs in each bin would be too small, and the relative importance of the linear displacements increases for decreasing bin size, causing noisy results. Consequently, the MSD is always computed in the entire bulk region (between $z = 40$ and 120). Since the MSD is already averaged over all runs, a statistical uncertainty as indicated in section 10.1 cannot be given.

The initial transient t_{tr} for the MSD computation at the beginning of the simulations was found to be about three inverse shear rates, but at least 2×10^4 time steps which is roughly the momentum diffusion time $t_{\mathrm{md}} = L_z^2/(8\nu_0) = 19200$ for a system of size $L_z = 160$ and viscosity $\nu_0 = \eta_0/\rho = \frac{1}{6}$. Therefore, the data within the initial time interval until $t_{\mathrm{tr}} = \max(3/\dot{\gamma}, t_{\mathrm{md}})$ is excluded from the analysis for the MSD. Neglecting a longer initial interval does not lead to significantly different results. These transients are shorter than for the collective ordering (section 10.6). This indicates that the MSD is dominated by local properties of the system.

Overall properties of the mean square displacements

Some exemplary MSD curves (soft RBCs at 65% volume fraction) for the particle motion along the y- and z-axes are shown in fig. 10.17. For the other simulation parameters, the MSDs are qualitatively equivalent and, therefore, are not shown. It can be seen that, similarly to molecular

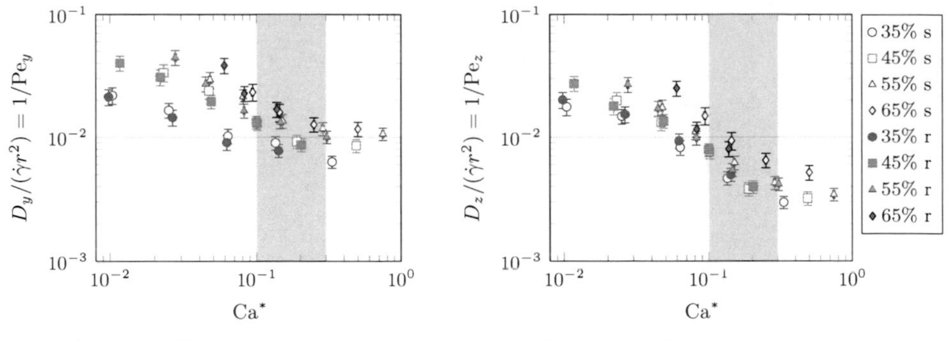

(a) reduced diffusivity along y-axis

(b) reduced diffusivity along z-axis

Fig. 10.18.: Reduced diffusivities of suspended red blood cells. The reduced diffusivities, (a) $D_y/(\dot{\gamma}r^2)$ and (b) $D_z/(\dot{\gamma}r^2)$, are shown as function of capillary number Ca*. They equal the inverse of the shear-induced Péclet numbers, Pe_α ($\alpha = y, z$), cf. eq. (10.26). The gray area denotes the region ($\mathrm{Ca}^* \in [0.1, 0.3]$) where tumbling is replaced by tank-treading.

hard sphere systems, a regime with a quadratic Δt-dependence is followed by a linear dependence at larger time shifts Δt. The transition sets in after about one inverse shear rate. At the beginning of the transition, the particles have moved by 0.1–0.2r which is of the order of the distance between particles. A plateau is absent for all investigated volume fractions and shear rates. This indicates that cage effects are unimportant for the studied range of parameters.

It must be emphasized that, for the small shear rates, the simulation time in terms of $\dot{\gamma}t$ is so short that the diffusive regime has only just developed. Longer simulations are necessary to improve the quality of the results in these cases. Still, they are satisfactory to perform some qualitative and quantitative investigations. The behavior of the MSD in the quadratic and the linear regimes is thoroughly analyzed in the following.

Linear regime and particle diffusion

In terms of displacements, diffusion along the y- and z-directions starts at MSD $\approx (0.3r)^2$ for all investigated cases. Since the MSD curves are mutually shifted along the Δt-axis (fig. 10.17), the onset of diffusion in terms of time shift $\dot{\gamma}\Delta t$ is varying. The reason for the shift is that the prefactor in the quadratic regime is a function of the applied shear rate as will be discussed below. The linear regime starts later if the particles are more deformed (after 1–5 inverse shear rates). Breedveld [94] has observed that the shear-induced diffusion regime for hard sphere systems starts at about $\dot{\gamma}\Delta t = 1$. This indicates that the deformability of the particles delays the onset of diffusion.

The diffusion coefficients D_y and D_z for the motion along the y- and z-axes are obtained by fitting $m\Delta t + n$ to the MSD curves in the Δt-interval where the curve is linear. Since the MSDs reflect the displacements in 1D, the gradient m is assumed to be twice the diffusivity. In the region where $\dot{\gamma}\Delta t$ is large, deviations from the linear behavior are observed. It is believed that these deviations are caused by the smaller number of available initial times over which the data can be averaged. Therefore, these regions are excluded from the fit. Longer simulation runs would reveal if this interpretation is correct. As mentioned above, the MSD data is directly averaged over all independent runs. Instead of the uncertainty based on statistical averaging, a relative ad-hoc error of 10% for the diffusivities is assumed. The intention is to keep track of the uncertainties and to distinguish physical effects from possible artifacts. The diffusivities have

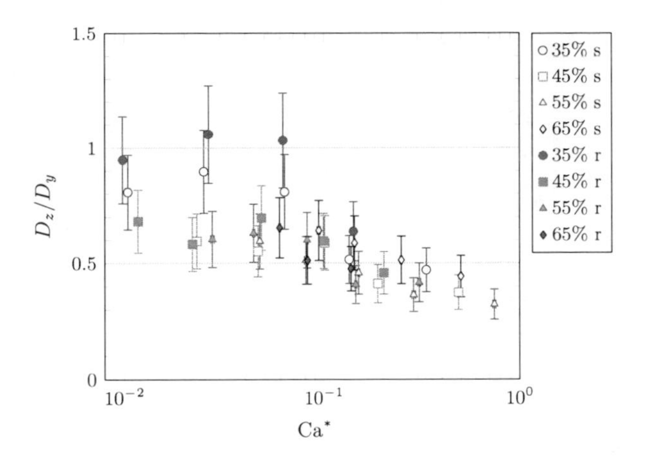

Fig. 10.19.: Relative importance of red blood cell diffusion along the y- and the z-axes. The data is shown as function of capillary number Ca*.

been found to be identical for both definitions of the centroid, eq. (10.22) and eq. (10.23). The reason is that the relative difference between both centroids is small compared to the particle displacements at larger Δt where the MSDs are linear.

For hard sphere systems, all components of the shear-induced diffusivity tensor scale like $D_{\alpha\beta} \propto \dot{\gamma}r^2$, and $\boldsymbol{D}$ is an increasing function of volume fraction up to 50% where it levels off [94]. In this region, typical values are $D_y/(\dot{\gamma}r^2) \approx 0.07$ (vorticity direction) and $D_z/(\dot{\gamma}r^2) \approx 0.11$ (velocity gradient direction). Fig. 10.18 contains the data for $D_y/(\dot{\gamma}r^2)$ and $D_z/(\dot{\gamma}r^2)$ as function of capillary number Ca*. The relative importance of D_z and D_y is shown in fig. 10.19. The diffusivity data turns out to be a rich source of information. The following list contains the most interesting and relevant observations.

1. The values for the diffusivities are substantially smaller than for hard sphere systems of comparable volume fractions.
2. Both reduced diffusivities, $\hat{D}_y := D_y/(\dot{\gamma}r^2)$ and $\hat{D}_z := D_z/(\dot{\gamma}r^2)$, are decreasing functions of the capillary number, and at least $\hat{D}_y$ seems to approach a plateau above $\mathrm{Ca}^*_{\mathrm{cr}} = 0.2$.
3. The diffusivity along the velocity gradient axis is nearly always smaller (down to a factor of $D_z/D_y = 0.3$) than that along the vorticity axis.
4. D_z/D_y is a slightly decreasing function of Ca*.
5. The reduced diffusivities $\hat{D}_y$ and $\hat{D}_z$ increase with the volume fraction for each value of Ca*.

The first two points indicate that the particle deformability slows down diffusive motion as compared to rigid spheres. When the RBCs are tank-treading ($\mathrm{Ca}^* > \mathrm{Ca}^*_{\mathrm{cr}}$), they seem to have a smaller tendency to collide, or their collisions are less effective in mixing the suspension. This fits into the picture that strongly deformed RBCs in shear flow tend to behave like isolated particles avoiding collisions with their neighbors.

The third observation is interesting because, in contrast to the obtained value $D_z/D_y \approx 0.3$–1, theory for rigid spheres [254] suggests $D_z/D_y \approx 1.5$. A similar result ($D_z/D_y \approx 1.7$) for volume fractions between 20 and 50% has been obtained experimentally [94]. There are essentially two possible reasons for this strong deviation from hard sphere systems. First, the particles are deformable and not spherical and have different extensions along different directions. Probably, not only the large radius r, but also the small radius $h/2$ plays a role for diffusion. Second, the

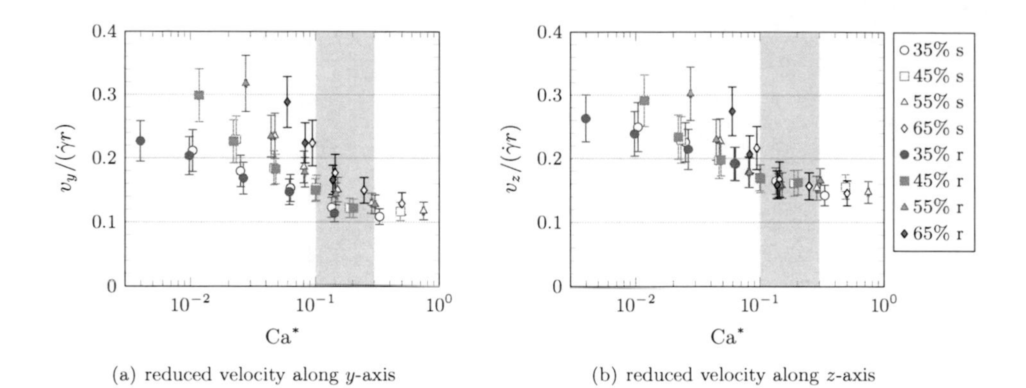

(a) reduced velocity along y-axis

(b) reduced velocity along z-axis

Fig. 10.20.: Reduced short time velocities of suspended red blood cells. The reduced velocities (a) $v_y/(\dot{\gamma}r)$ and (b) $v_z/(\dot{\gamma}r)$ are shown as function of the capillary number Ca*. The gray area denotes the region ($\text{Ca}^* \in [0.1, 0.3]$) where tumbling is replaced by tank-treading.

present system is bound by walls which may reduce the mobility of the particles along the z-axis. Additionally, shear-induced diffusion has been found to be sensitive to small periodic system sizes [255, 256]. It is possible that the system investigated in this work is still too small to find reasonable values for the diffusivities. This point should be taken into account in the future.

The fourth observation that D_z/D_y decreases slightly with Ca* may be caused by an increasing tendency of the tank-treading particles to form layers parallel to the xy-plane. Within these layers, particles may diffuse more easily than moving to other layers. This hypothesis should be tested in the future as well.

More frequent hydrodynamic collisions of the particles lead to stronger position fluctuations. However, experiments with hard spheres suggest that the reduced shear-induced diffusivities, $\hat{D} = D/(\dot{\gamma}r^2)$, increase only up to a volume fraction of about 50% [94]. This is related to the onset of crowding effects. In the present simulations, there is no indication for such a behavior (fifth observation).

It is not clear if there is a plateau for $\hat{D}_y$ and $\hat{D}_z$ at smaller values of Ca*. For rigid particles, the capillary number is not relevant, and $\hat{D}_y$ and $\hat{D}_z$ should not depend on it. Therefore, it would be interesting to investigate the diffusion at even smaller shear rates.

The Péclet number

$$\text{Pe}_\alpha := \frac{\dot{\gamma}r^2}{D_\alpha}, \quad (\alpha = y, z) \tag{10.26}$$

is a measure for the relative importance of particle advection and diffusion. It is the inverse of the reduced diffusivity. Typical Péclet numbers in the present simulations are of the order of 50–200. Therefore, advection within the shearing plane is significantly more important than diffusion in y- or z-direction. Typical thermal Péclet numbers for a RBC in shear flow ($r = 4\,\mu\text{m}$, $\dot{\gamma} = 100\,\text{s}^{-1}$, $D_{\text{th}} = \frac{kT}{6\pi\eta_0 r} \approx 5 \times 10^{-14}\,\text{m}^2\,\text{s}^{-1}$) are of the order of 3×10^4. Shear induced diffusion is more important, at least for shear rates above $1\,\text{s}^{-1}$. This provides an a posteriori justification why thermal fluctuations in the present simulations have not been taken into account. Contrarily, when RBCs are simulated at small shear rates ($< 1\,\text{s}^{-1}$), thermal fluctuations cannot be simply ignored.

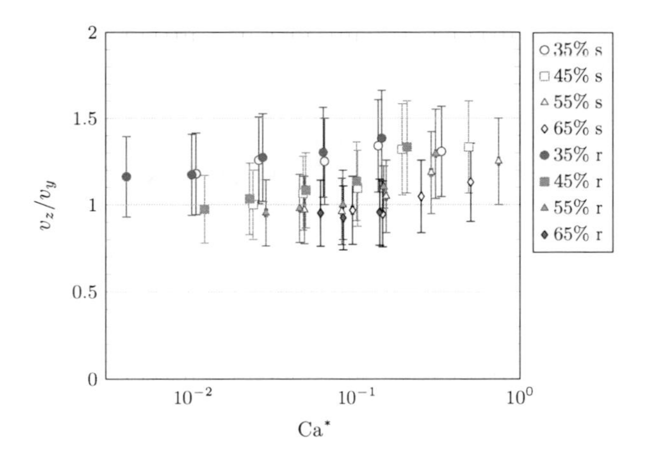

Fig. 10.21.: Relative importance of red blood cell short time velocities along the y- and the z-axes. The data is shown as function of the capillary number Ca*.

Quadratic regime and 'ballistic' particle motion

The 'ballistic velocity' of the RBCs at small displacements is found by fitting $v_\alpha^2 \Delta t^2$ ($\alpha = y, z$) to the MSD data for $\dot{\gamma}\Delta t \leq 0.15$. Within this interval, the MSDs have been found to be quadratic in Δt (fig. 10.17). Similarly to the diffusivities, the velocities v_y and v_z should be basically proportional to the applied shear rate $\dot{\gamma}$ because all motion in the system is shear-induced. Therefore, the reduced velocities $\hat{v}_y := v_y/(\dot{\gamma}r)$ and $\hat{v}_z := v_z/(\dot{\gamma}r)$ are shown in fig. 10.20. The data is obtained from taking the center of volume, eq. (10.23). For the less deformed particles (smaller Ca*), the results obtained from the center of surface, eq. (10.22), are practically identical. If, however, the particles are more strongly deformed (larger Ca*), deviations up to 8% in the linear velocities have been found (data not shown). Without exception, the velocities for the center of volume are larger than those for the center of surface. It is clear that the less deformed particles are generally more symmetric, leading to a smaller deviation of both center definitions. The deviations of 8% at larger Ca* do not significantly affect the qualitative discussion, and only the data obtained from the center of volume is considered in the following.

Important information can be extracted from fig. 10.20. Both $\hat{v}_y$ and $\hat{v}_z$ significantly decrease at $\mathrm{Ca}^* \approx 0.1$, i.e., when the particles start to tank-tread. Eventually, a plateau is reached. Above $\mathrm{Ca}^*_{\mathrm{tt}} = 0.2$, the velocities become independent of Ht. Below, higher volume fractions lead to larger velocities. This is in line with previous observations. Tank-treading particles basically behave like isolated objects which feel the other particles only via the suspension stress already contained in Ca*. For tumbling particles, higher volume fractions lead to stronger fluctuations in the ambient fluid which manifest themselves in the particle displacements.

The relative importance of the velocities v_z and v_y is shown in fig. 10.21. Both velocity components are equally important over the entire Ca*-range. Thus, the short time displacements in the yz-plane are nearly isotropic.

Caution is advised when the term 'ballistic' is used in the present work. The particles are immersed in a viscous fluid, and inertia effects are absent. Therefore, it is wrong to assume that particles just move with constant velocity until they touch a neighbor because hydrodynamic interactions are also present when the particles are isolated. However, this does not imply that the MSD cannot be quadratic in Δt at short time shifts. For rigid spheres immersed in a viscous fluid, Breedveld [94] even reports another linear regime (instead of a quadratic behavior) at small displacements ($\dot{\gamma}\Delta t < 0.1$), possibly due to the Brownian motion of the molecules of the

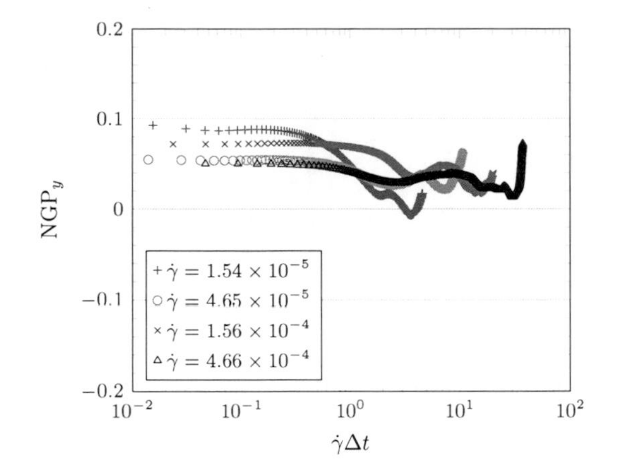

Fig. 10.22.: Non-Gaussian parameter (NGP) for diffusion of suspended red blood cells (RBCs). The data for the motion of the soft RBCs with Ht = 65% along the y-axis is shown. The NGP curves for the motion along the z-direction and the other simulations are qualitatively similar and are not shown.

suspending fluid. The mechanism responsible for the quadratic behavior seen in this section at time shifts $< 0.2\dot\gamma\Delta t$ (fig. 10.17) is not known, but it may be related to the deformation of the RBCs.

Non-Gaussianity

The displacement distributions for dense hard sphere systems are typically Gaussian in the ballistic and in the diffusive regimes, indicating uncorrelated particle motion. In the transitional region, when particles are caught in a cage of neighbors, motion may be correlated and the distributions may be non-Gaussian. One measure to investigate this property is the non-Gaussian parameter (NGP)

$$\mathrm{NGP} := \frac{M_4 - 3M_2^2}{M_2^2} \tag{10.27}$$

where

$$M_n := \int x^n G(x)\,\mathrm{d}x \tag{10.28}$$

is the n-th moment of the distribution $G(x)$ of variable x. If $G(x)$ is a Gaussian distribution, it obeys

$$G(x) = \frac{1}{\sqrt{2\pi\sigma^2}} \exp\left(\frac{(x-\bar{x})^2}{2\sigma^2}\right) \tag{10.29}$$

where $\bar{x}$ is the mean and σ^2 is the variance of the distribution of the variable x. The NGP is defined in such a way that it vanishes for a centered Gaussian ($\bar{x} = 0$).

In the present simulations, the transition between the quadratic and the linear regimes is found between $1/\dot\gamma$ and $5/\dot\gamma$. Even for the densest system, there is no sign of a plateau in the MSD curves (fig. 10.17). The NGPs for the displacements of the softer RBCs with 65% volume fraction are shown in fig. 10.22. Over the entire Δt-range, there is no significant deviation from zero. Rather, the NGP seems to be dominated by fluctuations caused by the finite size of the system and the reduced number of sample points at large values of Δt. These curves base on the center

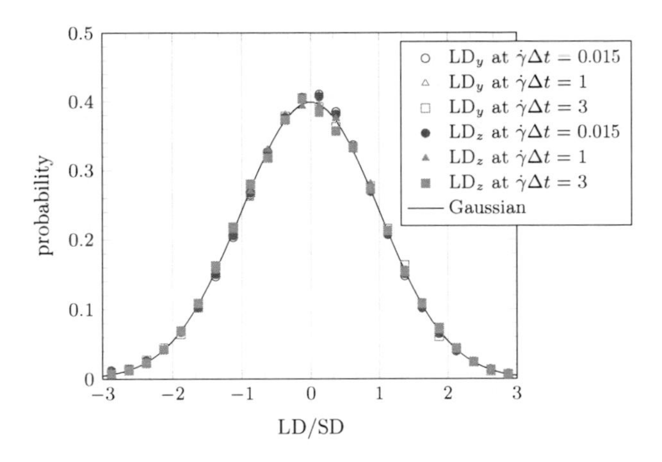

Fig. 10.23.: Displacement probability distributions of suspended red blood cells (RBCs). The probability distributions of the soft RBCs with Ht = 65% and the smallest shear rate ($\dot\gamma = 1.54 \times 10^{-5}$) for the linear displacements (LD) along the y- and z-axes are shown for three strains ($\dot\gamma\Delta t = 0.015$, 1, and 3). These strains belong to the quadratic, transitional, and linear regimes, respectively (fig. 10.17). The solid line denotes a Gaussian with unit standard deviation (SD) and zero mean. The other simulations lead to qualitatively similar results which are not shown.

of volume, eq. (10.23), but they are similar if eq. (10.22) is used instead. Fig. 10.23 shows three examples of the displacement distributions for the same suspension at the smallest shear rate (at $\dot\gamma\Delta t = 0.015$, 1, and 3). It can be inferred that there is no significant non-Gaussian property of the displacement distributions, neither in the quadratic, the linear, or in the transitional regime. The results for the other simulations are qualitatively similar and are not shown.

Concluding, the absence of a plateau in the MSD curves (fig. 10.17), the still increasing diffusivity with volume fraction in fig. 10.18, and the nearly vanishing NGPs in fig. 10.22 and fig. 10.23 strongly indicate that the present suspensions are far from the glassy state.

10.8. Shear stress fluctuations

The method of planes (section 9.4) allows of the computation of the particle stress averaged over planes parallel to the confining walls. Due to the microscopically inhomogeneous suspension structure, the local stresses are subject to permanent fluctuations about their macroscopic ensemble averages, $\boldsymbol{\sigma}(z,t) = \langle\boldsymbol{\sigma}\rangle + \delta\boldsymbol{\sigma}(z,t)$. In the following, $\langle\boldsymbol{\sigma}\rangle$ denotes the macroscopic stress (averaged over the entire bulk volume, the steady state, and all independent runs), and $\delta\boldsymbol{\sigma}(z,t)$ denotes the instantaneous stress fluctuation (averaged over the xy-plane). It has to be noted that the stress fluctuations are tightly connected to the finite system extension in the xy-plane. Therefore, it is expected that all fluctuations decrease like $N^{-1/2}$ where N is the number of particles in the plane.

Characterization of the fluctuations

Representative examples of the time evolution of the particle stress fluctuations for $\dot\gamma \approx 1.5 \times 10^{-4}$ and midway between the confining walls are shown in fig. 10.24. These curves are of relevance for future investigations when the dissipation mechanisms in the suspension are studied in more detail. Stress buildup and relaxation may be correlated with local events, such as instantaneous

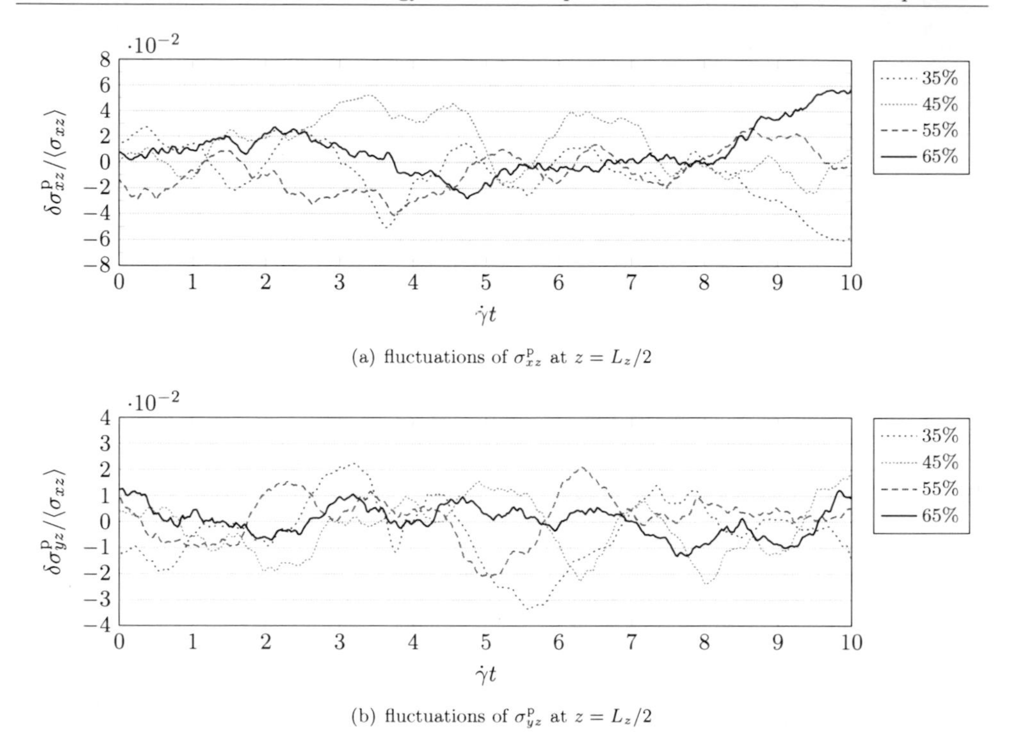

(a) fluctuations of σ^{P}_{xz} at $z = L_z/2$

(b) fluctuations of σ^{P}_{yz} at $z = L_z/2$

Fig. 10.24.: Time evolution of the particle stress fluctuations. The particle stress fluctuations (a) $\delta\sigma^{\mathrm{P}}_{xz}$ and (b) $\delta\sigma^{\mathrm{P}}_{yz}$ are shown as function of strain $\dot{\gamma}t$ for one representative simulation run for all investigated volume fractions. The data sets correspond to the stresses averaged over the xy-plane at $z = L_z/2$ (midway between the confining walls) for the soft red blood cell suspensions at shear rates $\dot{\gamma} \approx 1.5 \times 10^{-4}$ in steady state. The fluctuations are normalized by the average suspension stress $\langle\sigma_{xz}\rangle$. Typical relative fluctuations are of the order of 2% for the xz- and 1% for the yz-component.

particle rotation fluctuations, particle deformation, or non-affine displacements. Without the discussion in section 9.4, such an analysis would not be possible.

The standard deviations of the particle stress fluctuations (averaged over the bulk region, $z \in [40, 120]$) are shown in fig. 10.25. The error bars correspond to the statistical uncertainties related to the averaging over z. The first observation (fig. 10.25(a)) is that the standard deviations of $\delta\sigma^{\mathrm{P}}_{xz}$ are between 2 and 4% of the suspension stress $\langle\sigma_{xz}\rangle$ and do not significantly change over the entire Ca* range. Since all fluctuations in the present system are shear-induced, it is reasonable to normalize the fluctuations by the shear rate (and the constant viscosity η_0 to make the quantity dimensionless). The results are shown in fig. 10.25(b). In this picture, on the one hand, higher volume fractions lead to larger fluctuations. This is intuitively clear since a larger particle density should result in more significant distortions of the suspension. On the other hand, the fluctuations become less important for higher capillary numbers. This is particularly true in the region where tank-treading dominates, which is in line with the idea that tank-treading particles are more isolated and disturb the suspension less. A similar observation follows from fig. 10.25(d) where the fluctuations of σ^{P}_{yz}, normalized by $\eta_0\dot{\gamma}$, are shown. However, the dependence on Ca* is less pronounced than for the fluctuations of σ^{P}_{xz}. The fluctuations in the yz-plane, therefore, depend only slightly on the deformation state of the particles. Fig. 10.25(c), where the fluctuations of σ^{P}_{yz} are normalized by the stress $\langle\sigma_{xz}\rangle$, is hard to interpret. It seems to be counterintuitive that the relative fluctuations should increase with Ca* and decrease with Ht.

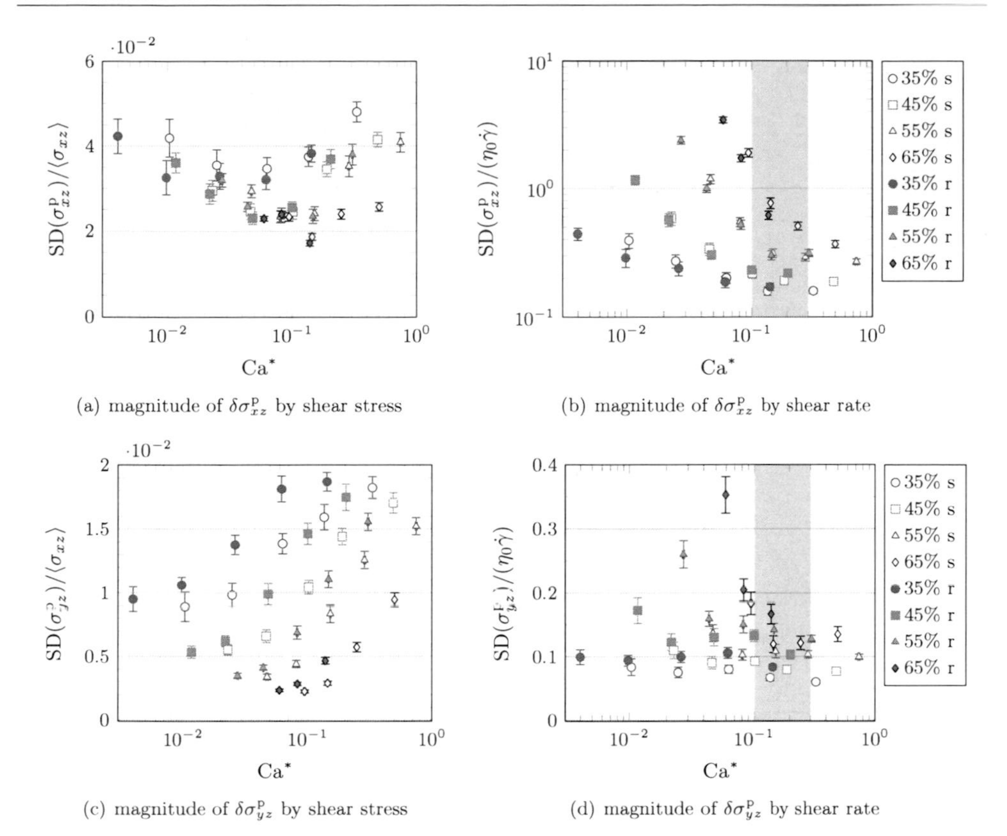

(a) magnitude of $\delta\sigma^{\mathrm{P}}_{xz}$ by shear stress

(b) magnitude of $\delta\sigma^{\mathrm{P}}_{xz}$ by shear rate

(c) magnitude of $\delta\sigma^{\mathrm{P}}_{yz}$ by shear stress

(d) magnitude of $\delta\sigma^{\mathrm{P}}_{yz}$ by shear rate

Fig. 10.25.: Relative magnitude of particle stress fluctuations. The standard deviations (SDs) of the xz- and the yz-components of the particle stress are shown. In (a) and (c), the SDs are normalized by the suspension stress $\langle\sigma_{xz}\rangle$, in (b) and (d) by the fluid stress $\eta_0\dot\gamma$. The gray area denotes the region ($\mathrm{Ca}^* \in [0.1, 0.3]$) where tumbling is replaced by tank-treading.

Therefore, the shear rate $\dot\gamma$, rather than the shear stress $\langle\sigma_{xz}\rangle$, appears to be responsible for the fluctuations in the yz-plane.

It has to be emphasized that the particle stress fluctuations, especially for the yz-component, depend on the z-position in the bulk region (data not shown). Therefore, wall effects are believed to be *not* negligible in the present discussion. The error bars in fig. 10.25 only describe the statistical uncertainty due to averaging over z, but a systematic deviation, caused by the presence of the walls, may be hidden. This may also explain why the data points for a given volume fraction in fig. 10.25(c) and fig. 10.25(d), unlike those in in fig. 10.25(a) and fig. 10.25(b), do not collapse. A larger system extension along the z-direction is required in order to give more reliable results.

Fig. 10.26 collects some examples of the particle stress distributions (both for $\delta\sigma^{\mathrm{P}}_{xz}$ and $\delta\sigma^{\mathrm{P}}_{yz}$). It is obvious that the distributions of $\delta\sigma^{\mathrm{P}}_{xz}$ are not Gaussian for small capillary numbers. This may be understood in the following way: Fig. 10.10 reveals that, for the densest suspension in the tumbling regime, $\bar\omega/\dot\gamma \approx 0.1$, i.e., the average RBC tumbling period is $T \approx 60/\dot\gamma$. This means that not the inverse shear rate is the largest time scale in the suspension. It is rather the rotation period of the particles. In order to obtain reasonable ensemble averages, at least half a rotation period should be simulated. This is roughly a factor five more than the simulation duration for the 65% suspension at the smallest shear rate. As a consequence, the steady state for the particle

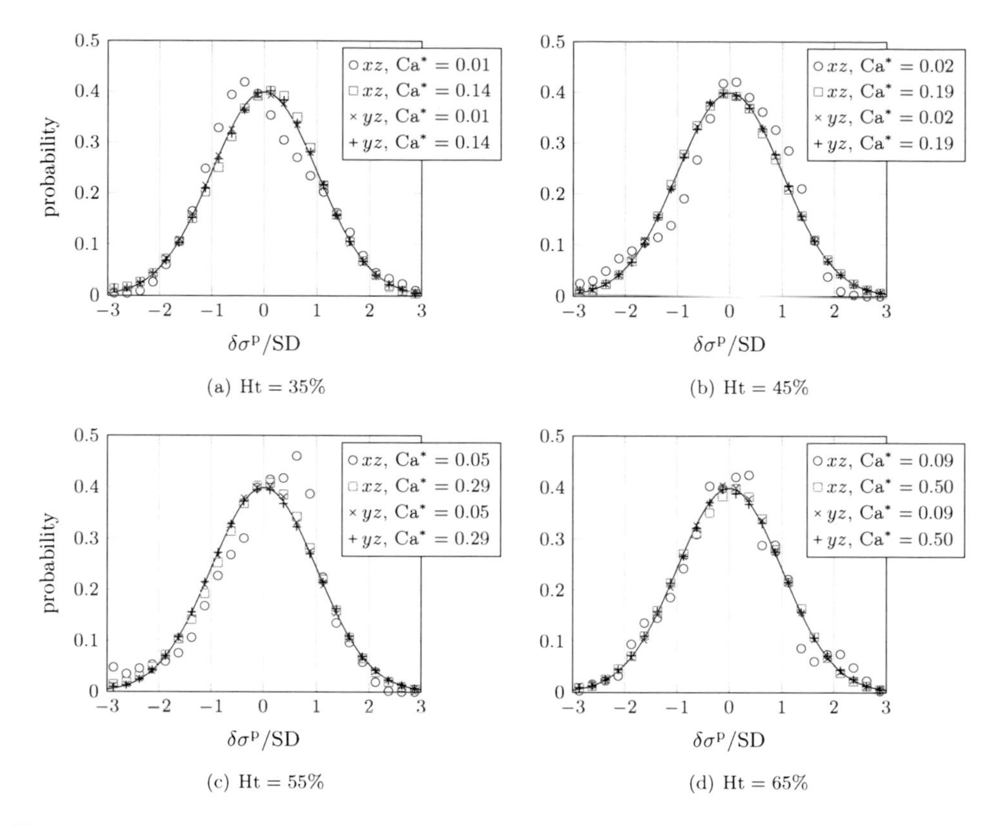

Fig. 10.26.: Distributions of particle stress fluctuations. The probability distributions of the xz- and yz-components of the particle stress are shown for the soft red blood cell suspensions for (a) 35%, (b) 45%, (c) 55%, and (d) 65% volume fraction and two different capillary numbers each (one in the tumbling, the other in the tank-treading regime). The fluctuations are normalized by their standard deviation (SD). All distributions, except for the xz-component at small capillary numbers, are Gaussian. A reference Gaussian distribution is shown as solid line. The data for the more rigid red blood cells is similar and is, therefore, not shown.

rotation has not been reached, independent runs are not equivalent, and ensemble averages are not yet well-defined. Contrarily, in the tank-treading state for larger values of Ca^*, tumbling rotation does not play a significant role, and the time scale is essentially set by the inverse shear rate. Thus, the ensemble average is sufficiently well-defined, and the particle stress fluctuations obey a Gaussian distribution. Interestingly, the distributions of the yz-component of the particle stress are always Gaussian, even for small Ca^*. The reason is that the RBC rotation about the x-axis is a pure fluctuation, $\langle \omega_x \rangle = 0$, and the corresponding transient is not set by the long rotation period T about the y-axis, but by the inverse shear rate $1/\dot{\gamma}$ which is shorter. Therefore, it may be assumed that the non-Gaussian shapes in fig. 10.26 are due to the comparably short simulation runtimes in terms of the strain $\dot{\gamma}t$. However, this question can only be answered by performing longer simulations in the future.

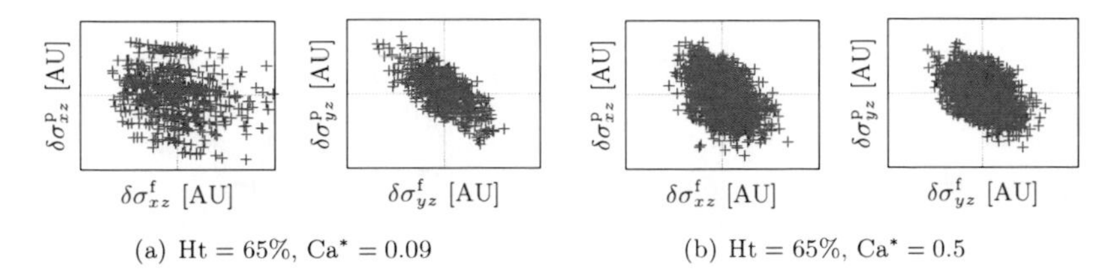

Fig. 10.27.: Correlation diagrams for the fluid and the particle stresses. Representative examples of the correlation diagrams for the soft red blood cell suspensions at Ht = 65% are shown. (a) Ca* = 0.09 (lowest shear rate) and (b) Ca* = 0.5 (highest shear rate). As the ensemble average $\langle \sigma^{\mathrm{p}}_{xz} \rangle$ is not well-defined for small shear rates, the correlation $\langle \delta\sigma^{\mathrm{f}}_{xz}, \delta\sigma^{\mathrm{p}}_{xz} \rangle$ in (a) is masked by noise. This leads to a decrease of the correlation in fig. 10.28 at small capillary numbers.

Correlations of shear rate and particle shear stress

The Pearson product-moment correlation coefficient of two functions $A(t)$ and $B(t)$ can be used to estimate the linear correlation between these functions. It is defined by

$$\langle A, B \rangle := \frac{\langle \delta A(t) \delta B(t) \rangle_t}{\sqrt{\langle \delta A^2(t) \rangle_t \langle \delta B^2(t) \rangle_t}} \in [-1, 1] \tag{10.30}$$

where $\delta A(t) := A(t) - \langle A(t) \rangle_t$ and $\delta B(t) := B(t) - \langle B(t) \rangle_t$ are the fluctuations of A and B. The time average is taken over the steady state. If the functions are linearly uncorrelated, $\langle A, B \rangle$ vanishes. A positive coefficient indicates correlation, a negative coefficient indicates anti-correlation.

The data obtained from the simulations is used to study the correlations between the shear rate and the particle shear stress, both averaged over the xy-plane, i.e., $A(t) = \langle \dot{\gamma} \rangle_{x,y}(z, t)$ and $B(t) = \langle \sigma^{\mathrm{p}}_{xz} \rangle_{x,y}(z, t)$. The resulting Pearson coefficient $\langle \dot{\gamma}, \sigma \rangle(z)$ is a function of z. It is then averaged over the bulk volume, $z \in [40, 120]$, and all independent runs. Fig. 10.27 shows representative examples of the correlation diagrams for $\langle \delta\sigma^{\mathrm{f}}_{xz}, \delta\sigma^{\mathrm{p}}_{xz} \rangle$ and $\langle \delta\sigma^{\mathrm{f}}_{yz}, \delta\sigma^{\mathrm{p}}_{yz} \rangle$. It can be seen that, in general, shear rate (i.e., fluid stress) and particle stress are anti-correlated. No cross-correlation such as $\langle \delta\sigma^{\mathrm{p}}_{xz}, \delta\sigma^{\mathrm{f}}_{yz} \rangle$ could be detected (data not shown). The (negative) Pearson coefficients for all simulations are shown in fig. 10.28. The shear rate and the shear stress are always anti-correlated. A local increase of the shear rate is related to a decrease of the particle stress. The degree of correlation strongly depends on the volume fraction and the capillary number. Denser systems are less correlated. For all systems except the densest one (Ht = 65%), the correlation is maximum for Ca* ≈ 0.1–0.2. This corresponds to the region where tank-treading sets in. For the system with 65% volume fraction, the correlation is not pronounced, but it steadily increases with Ca*.

The minor correlations for small values of Ca* in fig. 10.28 may be an artifact of an unsuitable ensemble average. As already discussed before, the simulation times for the smallest shear rates may be too short for a proper definition of the ensemble average of the particle stress. Inevitably, this would directly affect the definition of the stress fluctuation, and with it the definition of the Pearson coefficient. This becomes particularly visible in the left part of fig. 10.27(b): It seems that the data points in the scatter plot are arranged along parallel lines with negative gradient. Each single line would give rise to a large degree of anti-correlation. However, the entirety of the data points rather appears being more or less randomly scattered, which decreases the apparent correlation coefficient $\langle \delta\sigma^{\mathrm{f}}_{xz}, \delta\sigma^{\mathrm{p}}_{xz} \rangle$. Again, longer simulation runs seem to be the only reasonable option to improve the data statistics.

Still, fig. 10.28 supports the idea that stress release is tightly connected to the ability of the system to flow. It is expected that the stress increases when particles are locked. In this case,

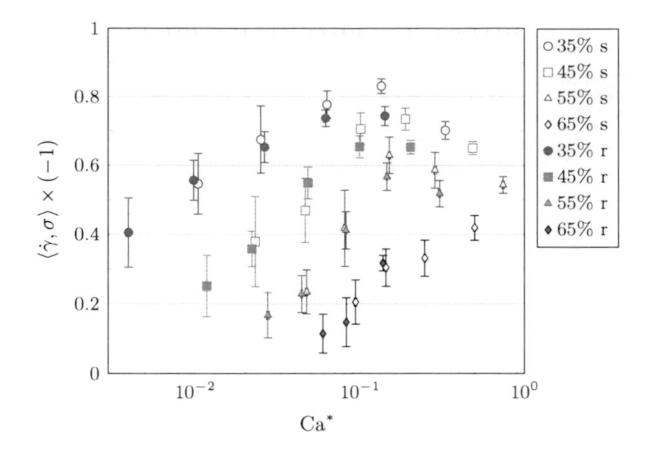

Fig. 10.28.: Pearson product-moment correlation coefficient for the shear rate $\dot{\gamma}$ and the particle shear stress σ^{p}_{xz}, cf. eq. (10.30). The correlation (multiplied by -1) is shown as function of capillary number Ca^*.

the local shear rate decreases because the particles cannot rotate and decelerate the ambient fluid. When the stress builds up, particles eventually pass each other, the shear rate increases again, and the stress relaxes, i.e., decreases. Although it is not directly clear how the correlation should depend on the capillary number Ca^*, it can be inferred that the particle deformation state and the transition from tumbling to tank-treading may play a role.

It has to be noted that the shear rate and the particle stress are first averaged over planes parallel to the walls. On average, depending on the volume fraction, about 20 particles are located on such a plane. It is therefore expected that correlations may be averaged out because some of the particles are in the act of being locked whereas others are just being freed. Ideally, the shear rate and shear stress in the neighborhood of each individual particle should be correlated before averaging. Unfortunately, the available data extracted from the simulations does not allow such an analysis. It is therefore proposed to correlate the instantaneous rotation state and the particle stress of individual particles in future investigations. Batchelor's approach (section 9.3) allows of the computation of the stress acting on each particle, whereas the instantaneous rotation state is more difficult to grasp (section 10.4). The particle deformability should also be taken into account during this analysis since energy can be stored elastically. It may also be rewarding to study the non-affine motion of particles relative to their neighborhood [257] and to correlate it with stress relaxation events. However, this is believed to be far from trivial. It is hardly imaginable that the particle center positions alone provide sufficient information about the connection of particle dynamics and stress relaxation.

11. Conclusions and outlook

This chapter is intended to give a concluding overview of the achievements of the present thesis. The contributions of this work to the scientific community and the physical results for the red blood cell simulations are reviewed in section 11.1. Open questions and suggestions for future research—both related to the physics and the code development—are pointed out in section 11.2.

11.1. Summary of own contributions and conclusions

During the course of preparing and writing this thesis, a series of new contributions to the skills of the research group in particular and to the knowledge of the community in general has been provided.

Own contributions

A computational model, based on the lattice Boltzmann method and the immersed boundary method, for the simulation of dense suspensions of deformable particles has been implemented and thoroughly analyzed (chapters 5, 6, 7, and 8). The algorithm is efficient in the sense that dense suspensions (65% volume fraction) with $\mathcal{O}(1000)$ particles can be simulated in a reasonable time (about 5 days for 10^5 time steps) on a modern single CPU with 3 GHz. The fluid viscosity, the applied shear rate or shear stress (see below), and the elastic particle properties can be directly controlled by the user. The numerical tool is not restricted to the simulation of red blood cells. Rather, a wide class of membrane-like particles for which the elastic constitutive law is provided can be simulated. It is hoped that the simulation tool will be further used in the future in order to study the rheology of dense suspensions of deformable particles (e.g., polydisperse capsules) or to investigate biomechanical processes relevant for medical research (such as platelet margination, cf. section 11.2).

In this context, some results regarding the numerical model (section 8.4) have already been published (Krüger et al. [187]). In this article, a single spherical, elastic capsule in simple shear has been simulated. The motivation for this investigation was to understand the reliability of the numerical method in the case of intermediate resolutions (particle radius about $8\Delta x$, Δx being the lattice constant). Higher resolutions are too expensive for the study of suspensions, and smaller resolutions do not permit high volume fractions. One of the major findings is that the hydrodynamic radius of the particles is slightly larger than the input radius ($r^* \approx r + 0.4\Delta x$, depending on the resolution and the interpolation stencil for the immersed boundary method). It is believed that the necessary interpolations between the Eulerian and the Lagrangian coordinate systems lead to the increase of apparent particle size. This has to be considered when results are compared with experiments or other simulations.

A shear stress boundary condition for the lattice Boltzmann method has been developed (section 5.4.2) and successfully tested (section 9.5). The primary importance of this boundary condition is that a shear flow can now be controlled by an applied shear *stress* rather than a shear *rate*. This is particularly important when the apparent viscosity of the fluid is not known as it is usually the case for complex fluids. Especially for future investigations of the yield stress in such systems, a shear stress boundary condition provides an interesting opportunity. As briefly described in section 2.3, one has to distinguish between the dynamic and the static yield stress of a system. The former is the shear stress which is found in the limit of vanishing but still *finite* shear rates. The latter is the stress below which the system deforms elastically and above which plastic deformations set in. The static is usually larger than the dynamic yield stress. It is now possible—within the lattice Boltzmann method—to identify both values independently. On the one hand, a fluid can be sheared with a small shear rate via standard velocity boundary conditions. This provides information about the dynamic yield stress. The new shear stress boundary condition, on the other hand, allows to access the static yield stress by increasing the load until the systems starts to deform plastically.

It is argued in appx. B.1.3 and shown via simulations in appx. B.1.4 that the deviatoric stress tensor in the lattice Boltzmann method is of second-order accuracy. These results have already been published (Krüger et al. [165, 258]). It is commonly claimed that boundary conditions for the lattice Boltzmann method should be designed in such a way that they maintain the second-order accuracy of the velocity field. The role of the stress tensor is usually neglected along the way, and its accuracy is affected. Therefore, it is hoped that those contributions help to improve the boundary conditions in such a way that they also retain the second-order convergence of the stress tensor.

Chap. 9 provides a coherent picture how stresses in the immersed boundary lattice Boltzmann method can be computed and evaluated. Especially the modified method of planes (section 9.4) turns out to be a useful tool for suspension rheology since it allows to compute the instantaneous and local[1] particle stress independently of the fluid stress (Krüger et al. [233]). Up to now, most researchers obtain the time-averaged particle stress from the total stress (which is known from macroscopic considerations for simple flow configurations, such as Poiseuille or simple shear flow) and the fluid stress computed within the lattice Boltzmann method. This approach, however, does not permit to access the spatio-temporal particle stress fluctuations.

The most important physical results and conclusions regarding the simulations of the red blood cell suspensions are collected in the following.

Conclusions from the simulations of the red blood cell suspensions

A major contribution to the understanding of the physical properties of red blood cell suspensions is provided in chap. 10 of the present work. For the first time, a systematic simulation study of the rheology of blood at intermediate shear rates is performed ($\dot{\gamma} \in [1, 100]\,\mathrm{s}^{-1}$ in physical units). The influence of the three major control parameters, the volume fraction Ht, the external shear rate $\dot{\gamma}$, and the particle rigidity κ_S, has been investigated.

It turns out that the three *input* control parameters (Ht, $\dot{\gamma}$, κ_S) are not best suitable for the description of the *outcome* of the simulations. Rather, the shear rate and the particle rigidity can be combined to a single parameter, the capillary number $\mathrm{Ca} := \eta_0 \dot{\gamma} r / \kappa_S$. Here, η_0 is the viscosity of the suspending fluid, and r is the large red blood cell radius. Data points for the viscosity and particle properties (such as deformation, rotation, alignment, and shear-induced diffusivity, see below) with the same values for Ht and Ca collapse, irrespective of the values for $\dot{\gamma}$ and κ_S. Thus, only two, rather than three parameters are required to characterize the simulation data. This behavior was to be expected from dimensional considerations.

Shear thinning behavior was observed for all combinations of control parameters (fig. 10.7). As expected, denser suspensions are more viscous, and the flow curves can be described by Ht and Ca alone. The flow curve for the suspension with 35% volume fraction was found to match experimental results at 45% [54] over two orders of magnitude in the shear rate (fig. 10.9). The reason for the apparent mismatch of the volume fractions is the effective hydrodynamic radius

[1] averaged over planes parallel to the walls

of the red blood cells as mentioned above. Taking the hydrodynamic radius into account, the results agree well.

In the case of dense suspensions, the 'corrected' capillary number $\text{Ca}^* := \eta\dot{\gamma}r/\kappa_S$ turns out to be more useful than Ca. It contains the total suspension stress $\eta\dot{\gamma}$, rather than the fluid stress $\eta_0\dot{\gamma}$. It is shown in chap. 10 that—under certain conditions—some of the data sets (red blood cell rotation, deformation, and alignment) can be described by Ca^* alone: Instead of the three input parameters (Ht, $\dot{\gamma}$, κ_S), Ca^* is sufficient to describe the observations! The reason for this astonishing behavior is the so-called 'tank-treading' of red blood cells. Thorough investigations revealed that tank-treading sets in at $\text{Ca}^*_{\text{cr}} \approx 0.2$ (fig. 10.10 and fig. 10.11). This is the case when the suspension shear stress becomes so large that the elastic stress of the membrane cannot maintain its characteristic biconcave shape, and the particle starts to rotate about its perimeter (like the track of a tank). When red blood cells are tank-treading, they behave nearly like isolated particles which are aware of the presence of their neighbors only via the suspension stress $\eta\dot{\gamma}$. This explains why even curves for different volume fractions collapse to a master curve for $\text{Ca}^* \geq \text{Ca}^*_{\text{cr}}$ when plotted as function of Ca^*.

This data collapse is especially important for four observables: the tumbling velocity (fig. 10.10), the deformation state (fig. 10.14), the inclination angle, and the orientational ordering (both in fig. 10.15) of the red blood cells. When the cells are tank-treading, their mutual interaction is reduced, and particle collisions become less important. There are several independent observations supporting this idea: (i) The average inclination angle of the red blood cells becomes a function of Ca^* alone and, on top, cannot be distinguished from that of an isolated cell under the same rheological conditions (fig. 10.15). (ii) The nematic order parameter dramatically increases at Ca^*_{cr} and reaches a plateau with the value ≈ 0.9. This indicates that the cells are strongly aligned (fig. 10.15). (iii) The average tumbling frequencies of the red blood cells decrease by more than one order of magnitude when the capillary number crosses the critical value 0.2. The interpretation is that the cells are in a nearly perfect tank-treading state without the need to tumble additionally. (iv) The average deformation state of the particles is a function of Ca^*, independent of the volume fraction. This is a hint that the suspension stress is the only relevant deformation mechanism for the red blood cells.

The reasons for shear thinning of blood commonly reported in the literature are the deformation (elongation), the tank-treading rotation, and the alignment of red blood cells [46]. In fact, the present simulations reveal that these three effects are tightly connected and not just independent mechanisms. The particle deformability eventually permits tank-treading when $\text{Ca}^* \geq \text{Ca}^*_{\text{cr}} \approx 0.2$, irrespective of volume fraction, shear rate, or particle rigidity. Due to tank-treading, the particles do not have to tumble any more and have the ability to align with their neighbors, increasing the orientational order. This way, layers of particles can slide over each other more easily These combined effects lead to a significant reduction of the suspension viscosity in the region $\text{Ca}^* \subset [0.1, 0.3]$ (fig. 10.12). However, also below the tank-treading regime, $\text{Ca}^* < \text{Ca}^*_{\text{cr}}$, the deformability leads to shear thinning behavior because particles can slightly give way and free themselves more efficiently when they are locked. The microscopic particle behavior (rotation and orientational alignment) differs strongly from that observed for a system of rigid ellipsoids [248]. This is a clear argument for the need of resolved *and* deformable red blood cells when hemorheology is studied via a bottom-up approach.

The present work shows, for the first time, a thorough investigation of red blood cell displacements for varying volume fraction, shear rate, and deformability in simple shear flow. MacMeccan [194] reported that the diffusive behavior was not obvious in his simulations, not even after 30 inverse shear rates. However, only 200 red blood cells have been used, and no additional independent runs have been performed. In contrast to this, it was possible, within the present simulations, to compute the shear-induced diffusivities of red blood cells along the vorticity and the velocity gradient directions (fig. 10.18). In dimensionless units, the diffusivities were found to be about one order of magnitude smaller as compared to sheared, non-Brownian hard sphere systems with comparable volume fractions [94]. The simulations reveal that the diffusion coefficients decrease with increasing Ca^*, especially in the regime where tank-treading sets in. This is additional evidence for the idea that tank-treading allows the particles to decouple their motion from their neighborhood to some extend. Collisions become less frequent, the particle motions are less distorted, and the diffusivity decreases. Still, the diffusivities increase with the volume fraction, supporting the expectation that a denser system leads to a larger number of particle collisions.

No clear sign of a yield stress could be found. The observations indicate that, in the present parameter space, even for volume fractions of 65%, glassy rheology plays no role. No plateau in the mean square displacements has been seen (fig. 10.17), the non-Gaussianity of the particle displacements does not significantly deviate from zero (fig. 10.22 and fig. 10.23), and the shear-induced diffusivity still seems to grow with the volume fraction (fig. 10.18). The flow curves (fig. 10.7) may lead to the assumption that a yield stress could be found for volume fractions larger than 45%. However, it is not clear how the viscosity behaves for $\dot{\gamma} \to 0$. Simulations with smaller shear rates are required.

It has also been found that the local shear rate and particle shear stress are correlated. The Pearson product-moment correlation coefficient is always negative with values between -0.1 and -0.8, depending on Ht and Ca^* (fig. 10.30). This means that a local increase of the shear rate leads to a local decrease of the stress. This may be interpreted in the following way: When red blood cells are locked during their rotation, they decelerate the ambient flow, and the shear rate decreases. At the same time, the stress builds up because the particle cannot move. After some time, the stress is large enough to push the particle out of its unfortunate situation. It rotates again, the shear rate increases, and the stress relaxes.

11.2. Outlook and suggestions for future research

In this section, the most relevant open questions arising from the present work are pointed out, and suggestions for future research are given. The list is divided into two parts: aspects of physical nature and issues related to possible extensions and improvements of the numerical model.

Physical aspects

The suspension behavior at small shear rates should be studied in more detail. It has been discussed in sections 10.3 and 11.1 that the present rheology data cannot provide clear evidence for the existence of a yield stress. Although the viscosity for volume fractions above 45% increases strongly when the shear rate is decreased, its fate in the limit of small shear rates is unclear. Additional simulations are required to distinguish between one of three possibilities: (i) existence of a Newtonian plateau, (ii) existence of a finite yield stress, (iii) continuing non-Newtonian properties in the absence of a yield stress. The second step would be the identification and understanding of the circumstances (e.g., volume fraction, particle shape, and deformability) under which each of these three possible behaviors can be found. Based on those insights, it may be possible to establish a proper rescaling for collapsing the flow curves. It can be inferred from fig. 10.7 that this rescaling must at least contain the volume fraction and the capillary number because the viscosity ratios depend on the capillary number. Later on, the similarities and differences to hard sphere suspensions may be worked out systematically. It has to be noted that simulations at small shear rates are expensive, which sets a lower bound to the accessible shear rates.

The suspension behavior in the large shear rate limit may be further investigated. In particular, it could be worked out for which capillary numbers a Newtonian plateau is fully developed and how its value depends on the microscopic details and the volume fraction. It is expected that higher spatial resolutions are required because the particle deformation becomes severe and large membrane curvatures are common. If the resolution is not sufficient, the simulations tend to become unstable.

Finite size effects cannot be completely ruled out in the present work. In principle, it should be tested which minimum system size (both along the directions with periodic boundary conditions and along the gradient direction between the walls) is required to obtain invariant statistical properties. Along this route, it would be important to check for the presence of a correlation length between, for example, the dynamics of particles and to test how this length depends on the shear rate and volume fraction. If this length becomes comparable to the size of the simulation box, the system size should be increased. It is expected that this type of correlation length, if ever, becomes important at high volume fractions only. Another source of finite size effects are hydrodynamic interactions which are well-known to be of long range. This type of finite size effect is, however, negligible at high volume fractions but should be considered when studying intermediate to low volume fractions. In all these cases, a careful study of system size-dependence is compulsory.

Another important question is how individual particle dynamics (e.g., instantaneous rotation, deformation, or non-affine motion) correlate with stress relaxation events in the suspension. One possible mechanism for stress relaxation may be the rotation of two particles about each other after they have been locked. This should be visible as an anti-correlation of the fluctuation of the angular velocity of the particles and the time derivative of their stress: Whenever particle rotations are hindered, the stress may increase. When particles can rotate again after being locked, the stress may relax. It is reasonable to assume that the deformation state of the particles also changes during these events. To this end, a more detailed description of the particle deformation may be required. The approach based on the inertia tensor (section 10.2) seems to be too simplistic for this purpose. Including the individual energy contributions (strain and bending) stored in the membrane deformation may also provide additional information. A robust and unique definition of the instantaneous rotation velocity of the particles (both for tumbling and tank-treading) would be of advantage. This way, not only an average, but also rotation velocity distributions could be reported. Yet, as discussed in sections 10.1 and 10.4, it is not clear if a unique rotation velocity can be defined for a strongly deformable object at all. A point not addressed in this work is the spatial distribution of the particles and their displacements: Can, at least for a finite time interval, layers or shear bands be identified? Can the concept of radial distribution functions be extended to non-spherical and deformable particles, and does it carry relevant information? Is it possible to assess non-affine displacements of the particles which may be another mechanism for stress relaxation? Such an analysis is aggravated by the deformability of the particles: A collision of two particles may not be directly visible when only their centers are tracked.

It may also be rewarding to study the local energy dissipation and the bidirectional transfer from fluid kinetic to membrane elastic energy. It is not clear which significance the fluctuations of the elastic energy stored in the membranes has and how it is related to particle diffusion and stress relaxation.

The developed code can and should be used to simulate biological systems, such as flows with resolved red blood cells and platelets and their hydrodynamic interactions. The code has already been extended for the inclusion of multiple kinds of particles with different properties (e.g., shape, size, and elastic moduli), including platelets and polydisperse suspensions. Albeit of its critical importance for the human body, platelet margination[2] in the circulatory system is still not understood [52, 259, 260].

[2]Margination is the tendency of the platelets to move towards the blood vessel walls where they can seal tissue ruptures.

Technical aspects

Each of the simulations performed in chap. 10 required between one day and three weeks computing time, depending on the shear rate and the volume fraction of the particles. The total CPU time for all simulations was about 2500 days, i.e., seven years. It is unlikely to significantly increase the single machine efficiency in the future. Instead, in order to simulate larger systems for a longer number of time steps, parallel simulations are eventually required. The implementation of a parallel version of the simulation code is well advanced [261] and should be ready for use within a few months after the completion of this thesis.

Wall effects can be avoided completely by the use of Lees-Edwards boundary conditions [144, 145]. This would circumvent the problems related to the definition of a bulk region. The major drawback, however, is that these boundary conditions and the method of planes (section 9.4) are not compatible. Within the Lees-Edwards approach, there seems to be no direct local access to particle stress fluctuations. It may still be worth to study additional possibilities for stress evaluation in Lees-Edwards simulations because their advantages over wall-driven simulations are considerable.

Viscosity ratios other than unity and viscoelastic membrane properties may be implemented in the code as well (e.g., [43, 93, 203] and [88, 89, 183], respectively). These features could be used to model specific biophysical or industrial suspensions. With some additional effort, the membrane code could also be extended in such a way that mesh reconfigurations are possible (e.g., [83, 183]). This way, vesicles and even droplets could be simulated with the immersed boundary method. However, it has to be noted that for fundamental questions, such as the possible existence of a yield stress in dense systems of deformable particles, the simulation parameter space should initially be kept as small as possible to allow simpler interpretations of the observations.

Part IV.

Appendix

A. Conventions, abbreviations, and symbols

A.1. Conventions

Throughout this thesis, the following conventions are used:

- Tensors and vectors are denoted by bold symbols (e.g., $\boldsymbol{\sigma}$ or $\boldsymbol{u}$), unspecified tensor and vector components by Greek letters (e.g., $\sigma_{\alpha\beta}$ or u_γ), and specified components by x, y, or z (e.g., σ_{xz} or u_y).
- The Einstein sum convention is used for repeated coordinate indices.
- Partial derivatives are often denoted by ∂ (e.g., $\partial_t := \partial/\partial t$, $\nabla_x := \partial/\partial x$, or $\nabla_\alpha := \partial/\partial x_\alpha$).
- Averages are denoted by angular brackets. The quantity over which is averaged is shown as index. For example, $\langle\boldsymbol{\sigma}\rangle_t$ is the time average of the stress tensor and $\langle\dot{\gamma}\rangle_{x,y}$ is the average of the shear rate over the xy-plane.
- When quantities are given in lattice units, the density ρ, the lattice constant Δx, and the time step Δt are usually set to unity if not otherwise stated.
- In the simulation parts of this work, quantities are usually given in lattice units if not otherwise stated.

A.2. Abbreviations

The following abbreviations are used throughout this thesis:

1D	one-dimensional, one dimension
2D	two-dimensional, two dimensions
3D	three-dimensional, three dimensions
BB	bounce-back
BC	boundary condition
BGK	Bhatnagar-Gross-Krook
FEM	finite element method
IBM	immersed boundary method
LBE	lattice Boltzmann equation
LBGK	lattice BGK
LBM	lattice Boltzmann method
LD	linear displacement
LGCA	lattice gas cellular automata
MOP	method of planes
MSD	mean square displacement
NGP	non-Gaussian parameter
NSE	Navier-Stokes equation
RBC	red blood cell
SD	standard deviation
WSS	wall shear stress

A.3. Symbols

Repeating and important symbols are collected in the following table:

General symbols

$\boldsymbol{x}$, $\boldsymbol{X}$, $\boldsymbol{r}$	position
t	time
$\boldsymbol{N}$	area normal vector
$\boldsymbol{n}$	area unit normal vector
$\boldsymbol{I}$	identity matrix

Hydrodynamic and rheological symbols

$\boldsymbol{u}$	fluid velocity
$\boldsymbol{u}^{\mathrm{w}}$	wall velocity
p	scalar pressure
ρ	density
$\boldsymbol{\sigma}$	deviatoric stress tensor
$\boldsymbol{\sigma}^{\mathrm{f}}$	viscous fluid stress tensor
$\boldsymbol{\sigma}^{\mathrm{p}}$	particle stress tensor
$\Delta\boldsymbol{p}$	momentum exchange
η	dynamic shear viscosity of the suspension
η_0	dynamic shear viscosity of the suspending fluid
ν	kinematic shear viscosity
$\boldsymbol{S}$	shear rate tensor
$\dot{\gamma}$	scalar shear rate
$\boldsymbol{f}$	force density (force per volume)
σ_{y}	yield stress
D_{th}	thermal diffusivity
$\boldsymbol{D}$	shear-induced diffusivity tensor
D_y, D_z	shear-induced diffusivities for motion along the y- and z-axes

Lattice Boltzmann symbols

L_x, L_y, L_z	system size (arbitrary)
N_x, N_y, N_z	system size (integer)
f_i	populations
f_i^*	post-collision populations
f_i^{eq}	equilibrium populations
f_i^{neq}	non-equilibrium populations
f_i^F	lattice force
τ	relaxation time (dimensionless)
q	number of lattice velocities
$\boldsymbol{c}_i$	lattice velocity
c_{s}	lattice speed of sound
Δx	lattice constant
Δt	time step size
w_i	lattice weight
$\boldsymbol{Q}_i$	velocity tensor
Ω_i	collision operator
ε	expansion parameter

Membrane symbols

r	radius
A	area, surface
V	volume
$\dot{\boldsymbol{x}}$	membrane node velocity
ϵ	energy density
E	energy
$\tilde{\boldsymbol{f}}$	force density (force per area)
$\boldsymbol{F}$	force
$\boldsymbol{V}$	displacement vector
$\boldsymbol{D}$	displacement gradient tensor
λ_1, λ_2	displacement gradient tensor eigenvalues
I_1, I_2	displacement gradient tensor invariants
H	curvature (trace of curvature tensor)
$H^{(0)}$	spontaneous curvature
κ_S	strain modulus
κ_α	area dilation modulus
κ_B	bending modulus
κ_A	surface extension modulus
κ_V	volume extension modulus (osmotic modulus)
κ_{int}	interaction modulus
κ_{gl}	wall roughness modulus (glue modulus)
$\boldsymbol{T}$	inertia tensor
T_1, T_2, T_3	principal moments of inertia
a, b, c	semiaxes of inertia ellipsoid
$\hat{\boldsymbol{o}}$	orientation unit vector
D_a, D_a^{max}	asymmetry deformation parameter (with maximum probability)
$\bar{\omega}$	average tumbling velocity
T	rotation period

Collective ordering symbols

$\boldsymbol{Q}$	nematic order tensor
$Q_>$	nematic order parameter
$\boldsymbol{n}$	director
θ	director inclination angle

Dimensionless numbers

Re	Reynolds number
Ma	Mach number
Kn	Knudsen number
Pe	Péclet number
Wi	Weissenberg number
Ca	capillary number
Ca^*, Ca^*_{cr}	(critical) corrected capillary number
Ht, ϕ	hematocrit, volume fraction

Functions

$\delta(x)$, $\delta(\boldsymbol{x})$	1D, 3D Dirac delta distribution
$\delta_\Delta(\boldsymbol{x})$	3D IBM interpolation stencil
$\phi(x)$	1D IBM interpolation stencil

B. Chapman-Enskog analysis and advanced lattice Boltzmann calculations

B.1. Chapman-Enskog analysis

The Chapman-Enskog analysis [262, 263, 264] has originally been used to derive the Navier-Stokes equations (NSE) and its transport coefficients from the Boltzmann equation [6, 118], but it can also be applied to the analysis of the lattice Bhatnagar-Gross-Krook (LBGK) equation [99]. The basic idea is that the fluid is assumed to be close to its local equilibrium everywhere. The magnitude of the deviation from equilibrium is controlled by the Knudsen number. For small deviations, a multi-scale expansion is possible, and it can be shown that the LBGK equation asymptotically solves the NSE where the density and pressure obey the equation of state of an ideal gas. In the following, the Chapman-Enskog analysis for the force-driven LBGK is presented. This analysis is restricted to a pure bulk system in the absence of any boundaries.

B.1.1. Chapman-Enskog analysis in the presence of forces

Multi-scale expansion

The Chapman-Enskog analysis bases on the assumption that the fluid is close to local equilibrium. For this reason, the lattice Boltzmann populations are expanded as

$$f_i = f_i^{(0)} + \epsilon f_i^{(1)} + \epsilon^2 f_i^{(2)} + \mathcal{O}(\epsilon^3) \tag{B.1}$$

where $f_i^{(0)} = f_i^{\mathrm{eq}}$, and $f_i^{(n)}$ ($n \geq 1$) is the n-th correction term. The expansion parameter, $\epsilon \ll 1$, can be identified as the Knudsen number [99]. Within the expansion in eq. (B.1), all coefficients are of the same order, i.e., $f_i^{(n)} = \mathcal{O}(1)$ for $n \geq 0$. The successively decreasing magnitude of the terms $\epsilon^n f_i^{(n)}$ is completely contained in ϵ^n.

The idea behind the multi-scale analysis is that there are two time scales in the LBM. The faster time scale is related to wave propagation, whereas the slower scale corresponds to momentum and mass diffusion [182]. For this reason, the time derivative $\partial_t := \partial/\partial t$ is also expanded,

$$\partial_t = \epsilon \partial_t^{(1)} + \epsilon^2 \partial_t^{(2)}. \tag{B.2}$$

Since the spatial variations of diffusion and advection are of the same order, the gradient is not decomposed and $\nabla = \epsilon \nabla^{(1)}$.

The Taylor expanded LBGK equation, eq. (5.3), takes the form

$$\sum_{n=1}^{\infty} \frac{1}{n!} (D_i \Delta t)^n f_i(\boldsymbol{x}, t) = -\frac{1}{\tau} \left(f_i(\boldsymbol{x}, t) - f_i^{(0)}(\boldsymbol{x}, t) \right) + f_i^F(\boldsymbol{x}, t) \Delta t \tag{B.3}$$

where $D_i := \partial_t + \boldsymbol{c}_i \cdot \nabla$. The ansatz for the lattice force is [110, 124]

$$f_i^F = w_i \left(A + \frac{\boldsymbol{B} \cdot \boldsymbol{c}_i}{c_s^2} + \frac{\boldsymbol{C} : \boldsymbol{Q}_i}{2 c_s^4} \right). \tag{B.4}$$

The parameters A, $\boldsymbol{B}$, and $\boldsymbol{C}$ are functions of the force density $\boldsymbol{f}$ and obey

$$\sum_i f_i^F = A, \quad \sum_i f_i^F \boldsymbol{c}_i = \boldsymbol{B}, \quad \sum_i f_i^F \boldsymbol{c}_i \boldsymbol{c}_i = c_s^2 A \boldsymbol{I} + \frac{1}{2}(\boldsymbol{C} + \boldsymbol{C}^{\mathrm{T}}). \tag{B.5}$$

Their values have to be obtained in the following. The force density is added on the ϵ level in the form $f_i^F = \epsilon f_i^{F(1)}$, $A = \epsilon A^{(1)}$, $\boldsymbol{B} = \epsilon \boldsymbol{B}^{(1)}$, and $\boldsymbol{C} = \epsilon \boldsymbol{C}^{(1)}$.

Inserting the above expansions into eq. (B.3), the resulting equation is sorted according to powers of ϵ. The coefficient equations for each power of ϵ have to be satisfied independently. The ϵ equation reads

$$D_i^{(1)} f_i^{(0)} = -\frac{1}{\tau \Delta t} f_i^{(1)} + f_i^{F(1)}, \tag{B.6}$$

and the ϵ^2 equation can be shown to be

$$\partial_t^{(2)} f_i^{(0)} + \left(1 - \frac{1}{2\tau}\right) D_i^{(1)} f_i^{(1)} = -\frac{1}{\tau \Delta t} f_i^{(2)} - \frac{\Delta t}{2} D_i^{(1)} f_i^{F(1)} \tag{B.7}$$

where $D_i^{(1)} := \partial_t^{(1)} + \boldsymbol{c}_i \cdot \nabla^{(1)}$.

In the absence of forces [110],

$$\sum_i f_i^{(0)} = \rho, \quad \sum_i f_i^{(0)} \boldsymbol{c}_i = \rho \boldsymbol{u} \tag{B.8}$$

and

$$\sum_i f_i^{(n)} = 0, \quad \sum_i f_i^{(n)} \boldsymbol{c}_i = 0 \tag{B.9}$$

for all $n > 0$, i.e., the correction terms do not contribute to mass density or momentum. However, due to discrete lattice effects [124], the velocity has to be redefined ($\boldsymbol{u} \to \boldsymbol{u}'$) if a force density $\boldsymbol{f}$ is included:

$$\sum_i f_i \boldsymbol{c}_i + m \boldsymbol{f} \Delta t = \rho \boldsymbol{u}' \tag{B.10}$$

where m is an a priori unknown constant. The correction term is added on the ϵ level,

$$\sum_i f_i^{(1)} \boldsymbol{c}_i = -m \boldsymbol{f}^{(1)} \Delta t. \tag{B.11}$$

i.e.,

$$\sum_i f_i^{(0)} \boldsymbol{c}_i = \rho \boldsymbol{u}' \tag{B.12}$$

holds where $f_i^{(0)} = f_i^{\mathrm{eq}}(\rho, \boldsymbol{u}')$. From here on, in order to avoid confusion, the dash will be dropped and $\boldsymbol{u}$ will be used again.

The following analysis will be performed in parallel for two different equilibrium populations, the standard quadratic form,

$$f_i^{(0)} = w_i \rho \left(1 + \frac{\boldsymbol{c}_i \cdot \boldsymbol{u}}{c_s^2} + \frac{\boldsymbol{Q}_i : \boldsymbol{u}\boldsymbol{u}}{2c_s^4}\right), \tag{B.13}$$

denoted (Q), and the linearized form,

$$f_i^{(0)} = w_i \rho \left(1 + \frac{\boldsymbol{c}_i \cdot \boldsymbol{u}}{c_s^2}\right), \tag{B.14}$$

denoted (L).

Euler equation

Based on the above considerations, the first two moments of eq. (B.6) read

$$\partial_t^{(1)}\rho + \nabla^{(1)} \cdot (\rho \boldsymbol{u}) = A^{(1)}, \tag{B.15}$$

$$\partial_t^{(1)}(\rho \boldsymbol{u}) + \nabla^{(1)} \cdot \boldsymbol{\Pi}^{(0)} = \left(\tilde{m} + \frac{m}{\tau}\right) \boldsymbol{f}^{(1)} \tag{B.16}$$

where the ansatz $\boldsymbol{B}^{(1)} = \tilde{m}\boldsymbol{f}^{(1)}$ is used ($\tilde{m}$ has to be determined) and

$$\boldsymbol{\Pi}^{(0)} = \sum_i f_i^{(0)} \boldsymbol{c}_i \boldsymbol{c}_i = \begin{cases} c_s^2 \rho \boldsymbol{I} + \rho \boldsymbol{u}\boldsymbol{u} & (Q) \\ c_s^2 \rho \boldsymbol{I} & (L) \end{cases}. \tag{B.17}$$

Eq. (B.17) is inferred from the isotropy relations, eq. (5.8). It can be seen that the Euler equation is recovered from eq. (B.15) and eq. (B.16) if $A^{(1)} = 0$ (continuity) and $\tilde{m} + m/\tau = 1$ (momentum balance). Obviously, the macroscopic pressure and the fluid density are connected via

$$p = c_s^2 \rho \tag{B.18}$$

which is the equation of state for an ideal gas. Therefore, eq. (B.15) and eq. (B.16) do not exactly describe an incompressible fluid.

Concluding, the ϵ equations asymptotically lead to the continuity and Euler equations on the $t^{(1)}$ time scale where the quadratic equilibrium gives rise to the non-linear advection term $\nabla^{(1)} \cdot (\rho \boldsymbol{u}\boldsymbol{u})$.

Navier-Stokes equation

In the following, ϵ is dropped from the equations by absorbing it back into the corresponding quantities, e.g., $\epsilon f_i^{(1)} \to f_i^{(1)}$ or $\epsilon \partial_t^{(1)} \to \partial_t^{(1)}$. This is not problematic since the separation of the scales has already been completed and the resulting equations, (B.6) and (B.7), have been obtained. In this sense, ϵ has merely been used as a tag for the different scales.

The first two moments of eq. (B.7) read

$$\partial_t^{(2)}\rho = \left(m - \frac{1}{2}\right) \nabla^{(1)} \cdot \boldsymbol{f}^{(1)} \Delta t, \tag{B.19}$$

$$\partial_t^{(2)}(\rho \boldsymbol{u}) = \left(m - \frac{1}{2}\right) \partial_t^{(1)} \boldsymbol{f}^{(1)} \Delta t + \nabla^{(1)} \cdot \boldsymbol{\sigma}^{(1)} \tag{B.20}$$

where the tensor $\boldsymbol{\sigma}^{(1)}$ is defined as

$$\boldsymbol{\sigma}^{(1)} = -\left(1 - \frac{1}{2\tau}\right) \boldsymbol{\Pi}^{(1)} - \frac{\Delta t}{4} \left(\boldsymbol{C}^{(1)} + \boldsymbol{C}^{(1)\mathrm{T}}\right) \tag{B.21}$$

and

$$\boldsymbol{\Pi}^{(1)} := \sum_i f_i^{(1)} \boldsymbol{c}_i \boldsymbol{c}_i = -\tau \Delta t \left(\partial_t^{(1)} \boldsymbol{\Pi}^{(0)} + \nabla^{(1)} \cdot \sum_i f_i^{(0)} \boldsymbol{c}_i \boldsymbol{c}_i \boldsymbol{c}_i - \frac{1}{2} \left(\boldsymbol{C}^{(1)} + \boldsymbol{C}^{(1)\mathrm{T}}\right) \right). \tag{B.22}$$

The last equality in eq. (B.22) follows from the second moment of eq. (B.6). As can be inferred from eq. (B.19) and eq. (B.20), the spatial and temporal derivatives of the force density $\boldsymbol{f}^{(1)}$ affect the density and the momentum on the $t^{(2)}$ time scale. This unphysical behavior can be eliminated by setting $m = \frac{1}{2}$ and $\tilde{m} = 1 - 1/(2\tau)$. The second term on the right-hand-side of eq. (B.22) can be written as

$$\nabla_\gamma^{(1)} \sum_i f_i^{(0)} c_{i\alpha} c_{i\beta} c_{i\gamma} = c_s^2 \left(\nabla_\alpha^{(1)}(\rho u_\beta) + \nabla_\beta^{(1)}(\rho u_\alpha) + \delta_{\alpha\beta} \nabla_\gamma^{(1)}(\rho u_\gamma) \right) \tag{B.23}$$

in component notation. Eq. (B.23) is valid both for the linear and the quadratic equilibria. Using eq. (B.15),

$$\partial_t^{(1)}\left(c_s^2\rho\delta_{\alpha\beta}\right) = -c_s^2\delta_{\alpha\beta}\nabla_\gamma^{(1)}(\rho u_\gamma), \tag{B.24}$$

the tensor $\mathbf{\Pi}^{(1)}$ can be simplified,

$$\Pi_{\alpha\beta}^{(1)} = \begin{cases} -\tau\Delta t\left(\partial_t^{(1)}(\rho u_\alpha u_\beta) + c_s^2\nabla_\alpha^{(1)}(\rho u_\beta) + c_s^2\nabla_\beta^{(1)}(\rho u_\alpha) - \frac{1}{2}\left(C_{\alpha\beta}^{(1)} + C_{\beta\alpha}^{(1)}\right)\right) & (Q) \\ -\tau\Delta t\left(c_s^2\nabla_\alpha^{(1)}(\rho u_\beta) + c_s^2\nabla_\beta^{(1)}(\rho u_\alpha) - \frac{1}{2}\left(C_{\alpha\beta}^{(1)} + C_{\beta\alpha}^{(1)}\right)\right) & (L) \end{cases}. \tag{B.25}$$

The remaining time derivative in eq. (B.25) for the quadratic equilibrium can be replaced by spatial derivatives by applying eq. (B.15) and eq. (B.16),

$$\begin{aligned} \partial_t^{(1)}(\rho u_\alpha u_\beta) &= u_\alpha u_\beta\nabla_\gamma^{(1)}(\rho u_\gamma) + u_\beta f_\alpha^{(1)} + u_\alpha f_\beta^{(1)} \\ &\quad - c_s^2 u_\beta\nabla_\alpha^{(1)}\rho - c_s^2 u_\alpha\nabla_\beta^{(1)}\rho - u_\beta\nabla_\gamma^{(1)}(\rho u_\alpha u_\gamma) - u_\alpha\nabla_\gamma^{(1)}(\rho u_\beta u_\gamma). \end{aligned} \tag{B.26}$$

Combining the previous results, one obtains

$$\Pi_{\alpha\beta}^{(1)} = \delta\Pi_{\alpha\beta}^{(1)} - \begin{cases} \tau\Delta t\left(c_s^2\rho\left(\nabla_\alpha^{(1)}u_\beta + \nabla_\beta^{(1)}u_\alpha\right) + \left(u_\alpha f_\beta^{(1)} + u_\beta f_\alpha^{(1)}\right) - \frac{1}{2}\left(C_{\alpha\beta}^{(1)} + C_{\beta\alpha}^{(1)}\right)\right) & (Q) \\ \tau\Delta t\left(c_s^2\rho\left(\nabla_\alpha^{(1)}u_\beta + \nabla_\beta^{(1)}u_\alpha\right) - \frac{1}{2}\left(C_{\alpha\beta}^{(1)} + C_{\beta\alpha}^{(1)}\right)\right) & (L) \end{cases}. \tag{B.27}$$

Expressions which involve derivatives of the density or which are of higher order in the velocity are contained in the error term

$$\delta\Pi_{\alpha\beta}^{(1)} = \begin{cases} -\tau\Delta t\nabla_\gamma^{(1)}(\rho u_\alpha u_\beta u_\gamma) & (Q) \\ -\tau\Delta t c_s^2\left(u_\beta\nabla_\alpha^{(1)}\rho + u_\alpha\nabla_\beta^{(1)}\rho\right) & (L) \end{cases}. \tag{B.28}$$

These error terms shall be ignored for now. They will be discussed again in appx. B.1.3.
With the help of eq. (B.27), eq. (B.21) can be written as

$$\sigma_{\alpha\beta}^{(1)} = \begin{cases} \left(\tau - \frac{1}{2}\right)c_s^2\Delta t\rho\left(\nabla_\alpha^{(1)}u_\beta + \nabla_\beta^{(1)}u_\alpha\right) + \Delta t\left(\left(\tau - \frac{1}{2}\right)\left(u_\alpha f_\beta^{(1)} + u_\beta f_\alpha^{(1)}\right) - \frac{\tau}{2}\left(C_{\alpha\beta}^{(1)} + C_{\beta\alpha}^{(1)}\right)\right) & (Q) \\ \left(\tau - \frac{1}{2}\right)c_s^2\Delta t\rho\left(\nabla_\alpha^{(1)}u_\beta + \nabla_\beta^{(1)}u_\alpha\right) + \frac{\tau\Delta t}{2}\left(C_{\alpha\beta}^{(1)} + C_{\beta\alpha}^{(1)}\right) & (L) \end{cases} \tag{B.29}$$

Interpreting

$$\nu = \left(\tau - \frac{1}{2}\right)c_s^2\Delta t \tag{B.30}$$

as the kinematic viscosity of the fluid and setting

$$C_{\alpha\beta}^{(1)} = \begin{cases} \left(1 - \frac{1}{2\tau}\right)\left(u_\alpha f_\beta^{(1)} + u_\beta f_\alpha^{(1)}\right) & (Q) \\ 0 & (L) \end{cases}, \tag{B.31}$$

$\boldsymbol{\sigma}^{(1)}$ can be identified as the deviatoric stress tensor for a viscous and incompressible Newtonian fluid,

$$\sigma_{\alpha\beta} = \rho\nu\left(\partial_\alpha u_\beta + \partial_\beta u_\alpha\right). \tag{B.32}$$

It can be obtained locally (i.e., without evaluating derivatives) from

$$\boldsymbol{\sigma}^{(1)} = \begin{cases} -\left(1 - \frac{1}{2\tau}\right) \boldsymbol{\Pi}^{(1)} - \frac{\Delta t}{2}\left(1 - \frac{1}{2\tau}\right)\left(\boldsymbol{u}\boldsymbol{f}^{(1)} + \boldsymbol{f}^{(1)}\boldsymbol{u}\right) & (Q) \\ -\left(1 - \frac{1}{2\tau}\right) \boldsymbol{\Pi}^{(1)} & (L) \end{cases}. \tag{B.33}$$

In practice, $\boldsymbol{\Pi}^{(1)}$ is replaced by $\boldsymbol{\Pi}^{\text{neq}} := \sum_i f_i^{\text{neq}} \boldsymbol{c}_i \boldsymbol{c}_i$ since the first-order correction $f_i^{(1)}$ is not known explicitly. This leads to an additional error which will be discussed in appx. B.1.3.

Concluding, the deviatoric stress tensor is recovered on the $t^{(2)}$ time scale. It is a non-equilibrium property of the fluid as it is encoded in the first-order non-equilibrium populations, $f_i^{(1)}$. It is also interesting to note the functional form of the viscosity in eq. (B.30): The additional term $\frac{1}{2}$ is caused by the spatial discretization of the LBM. It is called 'lattice viscosity' or 'propagation viscosity'. Fortunately, its contribution can be completely absorbed by redefining the physical viscosity, $\tau c_s^2 \Delta t \to \left(\tau - \frac{1}{2}\right) c_s^2 \Delta t$.

Combination of multi-scale results

The ϵ equations, eq. (B.15) and eq. (B.16), become

$$\partial_t^{(1)} \rho + \nabla^{(1)} \cdot (\rho \boldsymbol{u}) = 0, \quad \partial_t^{(1)} (\rho \boldsymbol{u}) + \nabla^{(1)} \cdot \boldsymbol{\Pi}^{(0)} = \boldsymbol{f}^{(1)}, \tag{B.34}$$

and the ϵ^2 equations, eq. (B.19) and eq. (B.20), are

$$\partial_t^{(2)} \rho = 0, \quad \partial_t^{(2)} (\rho \boldsymbol{u}) = \nabla^{(1)} \cdot \boldsymbol{\sigma}^{(1)}. \tag{B.35}$$

Combination of the ϵ and the ϵ^2 equations yields the continuity equation,

$$\partial_t \rho + \nabla \cdot (\rho \boldsymbol{u}) = 0, \tag{B.36}$$

and the momentum equation,

$$\partial_t (\rho \boldsymbol{u}) = \begin{cases} -\nabla \cdot (\rho \boldsymbol{u}\boldsymbol{u}) - \nabla p \boldsymbol{I} + \nabla \cdot \sigma + \boldsymbol{f} & (Q) \\ -\nabla p \boldsymbol{I} + \nabla \cdot \sigma + \boldsymbol{f} & (L) \end{cases} \tag{B.37}$$

where $p = c_s^2 \rho$ holds. Obviously, the incompressible NSE are not exactly solved since the LBM describes a compressible fluid and ρ is generally not constant in space and time. A discussion of the error terms appearing when the LBGK algorithm is used to solve the *incompressible* NSE is given in appx. B.1.3.

B.1.2. Diffusive scaling and its relevance for the convergence of the LBGK algorithm

When the convergence behavior of the LBM is discussed, the so-called 'diffusive scaling' plays a major role. It states that, if the spatial resolution is refined, $\Delta x \to \Delta x' = q \Delta x$ with $q < 1$, the time step has to be changed according to $\Delta t \to \Delta t' = q^2 \Delta t$, i.e., $\Delta t \propto \Delta x^2$ during refinement. It can directly be inferred from eq. (5.43) that τ is constant under these circumstances. The term diffusive scaling has no physical meaning. Rather, its functional form resembles the diffusion equation. It should be noted that, in principle, any other scaling $\Delta t \propto \Delta x^r$ may be realized as long as eq. (5.43) remains valid. In the general case, the relaxation parameter scales like $\left(\tau - \frac{1}{2}\right) \propto \Delta x^{r-2}$.

There are generally three error sources in LBM simulations [126, 164]: (i) a spatial discretization error $\propto \Delta x^2$, (ii) a temporal discretization error $\propto \Delta t^2$, and (iii) a compressibility error $\propto$

Tab. B.1.: Behavior of observables in the diffusive scaling, $\Delta t \propto \Delta x^2$. This scaling is only valid for the lattice values of these quantities. The physical values remain the same, i.e., the diffusive scaling is a pure refinement scheme keeping all dimensional quantities untouched.

quantity	symbol	scales like
velocity	$\boldsymbol{u}$	Δx
length	l	$1/\Delta x$
time interval	t	$1/\Delta x^2$
force density (per volume)	$\boldsymbol{f}$	Δx^3
stress	$\boldsymbol{\sigma}$	Δx^2
pressure	p	Δx^2
density gradient	$\partial_\alpha \rho$	Δx^3
spatial derivative	∂_α	Δx
time derivative	∂_t	Δx^2
lattice Mach number	Ma	Δx
Reynolds number	Re	1
relaxation parameter	τ	1

$\Delta t^2/\Delta x^2$. The distinctiveness of the diffusive scaling is that all of these contributions are at least of second order in Δx, i.e., the total error scales like Δx^2 in the diffusive scaling. Therefore, the second-order accuracy of the LBM can be spoiled if a wrong scaling is used, e.g., if Δx is refined but $\Delta t \propto \Delta x$. In this case, the compressibility error would not decrease, and the solution would not converge to the incompressible Navier-Stokes solution. In other words: When the lattice is refined, one has to make sure that the compressibility error is also reduced so that asymptotic recovery of the NSE and the incompressible limit (Ma $\to 0$) are guaranteed at the same time. As a result, within this scaling, all error terms are at least of second order in Ma and first order in Δt [164, 165]. The disadvantage of the diffusive scaling is that the number of simulation time steps is quadrupled when the resolution is doubled. In 3D, the total runtime of a simulation thus increases by a factor of $32 = 2^5$ when the resolution is doubled! This is one of the major drawbacks of the LBM.

Since the density is not changed during rescaling, it is straightforward to identify the scaling for each quantity (in lattice units) based on its physical unit. The most important quantities and their behavior under the diffusive scaling are shown in tab. B.1. Additional information is also given in Krüger et al. [165].

B.1.3. Error terms of the LBGK equation and its convergence to the Navier-Stokes equations

LBGK convergence

It has been shown in appx. B.1.1 that the LBGK solves the macroscopic equations

$$\partial_t \rho + \nabla \cdot (\rho \boldsymbol{u}) = 0 \tag{B.38}$$

and

$$\begin{aligned} \partial_t(\rho \boldsymbol{u}) + \nabla \cdot (\rho \boldsymbol{u}\boldsymbol{u}) &= -\nabla p \boldsymbol{I} + \nabla \cdot \left(\rho\nu \left(\nabla \boldsymbol{u} + (\nabla \boldsymbol{u})^{\mathrm{T}}\right)\right) + \boldsymbol{f} + \delta \boldsymbol{M}, \quad (Q) \\ \partial_t(\rho \boldsymbol{u}) &= -\nabla p \boldsymbol{I} + \nabla \cdot \left(\rho\nu \left(\nabla \boldsymbol{u} + (\nabla \boldsymbol{u})^{\mathrm{T}}\right)\right) + \boldsymbol{f} + \delta \boldsymbol{M}, \quad (L) \end{aligned} \tag{B.39}$$

where $p = c_s^2 \rho$ and $\nu = \left(\tau - \frac{1}{2}\right) c_s^2 \Delta t$ and contributions of order ϵ^3 and higher have been neglected. This is reasonable since the Knudsen number ϵ is proportional to the Mach number [265] and Ma $\propto \Delta x$ in the diffusive scaling. The additional error introduced by the deviatoric stress tensor reads $\delta \boldsymbol{M} = -(1 - 1/(2\tau)) \nabla \cdot \delta \boldsymbol{\Pi}^{(1)}$ where $\delta \boldsymbol{\Pi}^{(1)}$ is given in eq. (B.28). There are obviously two error sources in eq. (B.38) and eq. (B.39): (i) the error contained in $\delta \boldsymbol{M}$ and (ii) the error connected to the numerical compressibility of the fluid which appears via $\partial_t \rho$ and $\nabla \rho$. The compressibility errors are known to be of second order in the diffusive scaling [164, 265, 266], i.e., they scale with an additional factor Δx^2 compared to the dominating, physical terms. Additionally, the same holds for the error related to $\delta \boldsymbol{M}$: The terms given in eq. (B.28) scale like Δx^4, which can be inferred from tab. B.1. The gradient in $\delta \boldsymbol{M}$ introduces an additional order in Δx, and $\delta \boldsymbol{M} \propto \Delta x^5$. Since the dominating terms in eq. (B.39) scale like Δx^3, $\delta \boldsymbol{M}$ poses a second-order error.

Concluding, in the diffusive scaling, the macroscopic solution for the velocity of the LBGK equation converges to the NSE with a second-order rate, and relative errors decrease like $\Delta t \propto \Delta x^2 \propto \mathrm{Ma}^2$.

Stress tensor convergence

Although the NSE is asymptotically solved by the LBGK algorithm, the deviatoric stress tensor $\boldsymbol{\sigma}$ is a priori not known. The reason is that only its divergence enters the NSE and that $\boldsymbol{\sigma}$ does not have to be found explicitly in order to solve the LBGK equation. Fortunately, the stress tensor $\boldsymbol{\sigma}$ can be computed additionally as presented in eq. (B.33). The interesting property of the LBM is that this stress tensor is obtainable locally, i.e., without evaluating spatial derivatives.

From the discussion in appx. B.1.1, it could be inferred that the deviatoric stress tensor $\boldsymbol{\sigma}$ obtained from the non-equilibrium populations is not exactly the stress tensor expected for the *incompressible* NSE. In the previous paragraph it has already been discussed that the corresponding error term is of second order. However, an additional error $\delta \boldsymbol{\sigma}^*$ is introduced because the first order populations $f_i^{(1)}$ required for the evaluation of $\boldsymbol{\Pi}^{(1)}$ are not known directly. Instead, the known non-equilibrium populations, $f_i^{\mathrm{neq}} = \sum_{n \geq 1} f_i^{(n)}$, are taken to estimate the tensor $\boldsymbol{\Pi}^{(1)}$. Therefore, the contribution of the tensor $\boldsymbol{\Pi}^{(2)} := \sum_i f_i^{(2)} \boldsymbol{c}_i \boldsymbol{c}_i$ to $\boldsymbol{\sigma}$ should be assessed. Since each higher correction of the populations is smaller than the previous correction, $f_i^{(n+1)} \approx \epsilon f_i^{(n)} \ll f_i^{(n)}$, higher order tensors like $\boldsymbol{\Pi}^{(3)}$ are neglected and the additional error term can be approximated by

$$\delta \boldsymbol{\sigma}^* \approx -\left(1 - \frac{1}{2\tau}\right) \boldsymbol{\Pi}^{(2)}. \tag{B.40}$$

At this point, only the scaling and not the exact expression for the tensor $\boldsymbol{\Pi}^{(2)}$ is of interest: If $\boldsymbol{\Pi}^{(2)} \propto \Delta x^4$ is satisfied in the diffusive scaling, the LBGK can also be considered second-order accurate with respect to the deviatoric stress tensor $\boldsymbol{\sigma}$ since $\boldsymbol{\sigma} \propto \Delta x^2$.

In fact, it has already been shown analytically and numerically that the bulk stress tensor is indeed recovered with a second-order accuracy if (i) the quadratic equilibrium is employed, (ii) no external forces are included, and (iii) boundaries are absent (Krüger et al. [165, 258]). In the following, a sketch for the more general case will be provided: quadratic and linearized equilibria in the presence of forces.

The second moment of the ϵ^2 equation, eq. (B.7), reads (setting $\Delta t = 1$)

$$\begin{aligned} -\frac{1}{\tau} \boldsymbol{\Pi}^{(2)} = {} & \underbrace{\partial_t^{(2)} \boldsymbol{\Pi}^{(0)}}_{\text{term 1}} + \left(1 - \frac{1}{2\tau}\right) \underbrace{\partial_t^{(1)} \boldsymbol{\Pi}^{(1)}}_{\text{term 2}} + \left(1 - \frac{1}{2\tau}\right) \underbrace{\nabla^{(1)} \cdot \boldsymbol{R}^{(1)}}_{\text{term 3}} \\ & + \underbrace{\frac{1}{2} \partial_t^{(1)} \sum_i f_i^{F(1)} \boldsymbol{c}_i \boldsymbol{c}_i}_{\text{term 4}} + \underbrace{\frac{1}{2} \nabla^{(1)} \cdot \sum_i f_i^{F(1)} \boldsymbol{c}_i \boldsymbol{c}_i \boldsymbol{c}_i}_{\text{term 5}} \end{aligned} \tag{B.41}$$

where $\boldsymbol{R}^{(1)} := \sum_i f_i^{(1)} \boldsymbol{c}_i \boldsymbol{c}_i \boldsymbol{c}_i$. In the following, the scaling of all five terms is briefly discussed. If all of them scale at least $\propto \Delta x^4$ in the diffusive scaling, also the convergence of the stress tensor can be considered second-order.

1. From eq. (B.35) it follows that this term is $\propto \boldsymbol{u}\nabla^{(1)} \cdot \boldsymbol{\sigma}^{(1)} \propto \Delta x^4$. For the linearized equilibrium, it exactly vanishes.
2. Based on eq. (B.15), eq. (B.16), and eq. (B.27), one can infer that this term has contributions $\propto \left(\nabla^{(1)}\boldsymbol{u}\right) \nabla^{(1)} \cdot (\rho \boldsymbol{u})$, $\propto \rho \nabla^{(1)} \nabla^{(1)} \cdot \boldsymbol{\Pi}^{(0)}$, $\propto \rho \nabla^{(1)} \boldsymbol{f}^{(1)}$, and $\propto \rho \nabla^{(1)} \left(\boldsymbol{u}\nabla^{(1)} \cdot (\rho \boldsymbol{u})\right)$ which are all of order Δx^4.
3. The third moment of the populations $f_i^{(1)}$ can be computed from eq. (B.6),

$$\boldsymbol{R}^{(1)} = -\tau \partial_t^{(1)} \sum_i f_i^{(0)} \boldsymbol{c}_i \boldsymbol{c}_i \boldsymbol{c}_i - \tau \nabla^{(1)} \cdot \sum_i f_i^{(0)} \boldsymbol{c}_i \boldsymbol{c}_i \boldsymbol{c}_i \boldsymbol{c}_i + \tau \sum_i f_i^{F(1)} \boldsymbol{c}_i \boldsymbol{c}_i \boldsymbol{c}_i. \tag{B.42}$$

   The first term in eq. (B.42) scales $\propto \partial_t^{(1)} \rho \boldsymbol{u} = -\nabla^{(1)} \cdot \boldsymbol{\Pi}^{(0)} + \boldsymbol{f}^{(1)} \propto \Delta x^3$ and the second term $\propto \nabla^{(1)}(\rho \boldsymbol{u}\boldsymbol{u}) \propto \Delta x^3$. The last term has the same form as term 5 in eq. (B.41). Consequently, the divergence of $\boldsymbol{R}^{(1)}$ scales like Δx^4.
4. For the linearized equilibrium, this term vanishes. For the quadratic equilibrium, according to eq. (B.5) and eq. (B.31), it scales $\propto \partial_t^{(1)} \boldsymbol{C}^{(1)} \propto \partial_t^{(1)} \boldsymbol{u}\boldsymbol{f}^{(1)}$ which is at least of order Δx^4.
5. From the definition of f_i^F in eq. (B.4) it follows that this term is $\propto \nabla^{(1)} \boldsymbol{f}^{(1)}$ and therefore of order Δx^4.

Concluding, all error terms which are involved in the computation of the deviatoric stress tensor according to eq. (B.33) scale at least like Δx^4 whereas $\boldsymbol{\sigma}^{(1)}$ itself scales like Δx^2. For this reason, the deviatoric stress tensor in the bulk LBGK algorithm can be considered formally second-order accurate. In appx. B.1.4, based on results published in Krüger et al. [258], it is shown numerically that the stress tensor indeed is recovered with second-order convergence when the force-free and periodic Taylor-Green vortex is considered. A corresponding numerical investigation for a flow configuration subject to non-constant and inhomogeneous forcing has not been performed and is left for future research.

Finally, it must be emphasized again that this analysis is only valid for the bulk fluid without boundaries. Boundary conditions may introduce additional error terms for the stress tensor which scale more weakly than Δx^4 in the diffusive scaling. This is a priori not surprising since boundary conditions for the LBM are designed in such a way that they are of second-order convergence for the velocity, but the direct effect on the stress convergence is usually ignored. In Krüger et al. [165], it has been shown that the deviatoric stress tensor $\boldsymbol{\sigma}$ shows a better than first-order convergence rate for a flow in a square duct subject to bounce-back and velocity boundary conditions. Still, unambiguous second-order convergence could only be observed for the velocity, not for the deviatoric stress. Since second-order convergence of the velocity and the stress could be clearly recovered in a pure bulk system (appx. B.1.4 and Krüger et al. [258]), there is reason to assume that the presence of boundaries indeed may reduce the convergence order of the stress. This knowledge may be used in the future to design boundary conditions for the LBM which comprise a higher accuracy for the deviatoric stress tensor.

B.1.4. Benchmark test: Convergence for the Taylor-Green vortex flow

This section bases on the investigations published in Krüger et al. [258].

As stated in section 5.2 and in appx. B.1, the LBGK algorithm yields second-order accurate results both for the fluid velocity and the deviatoric stress in the diffusive scaling (section B.1.2): If the spatial resolution is increased (Δx reduced) and the time step is refined according to $\Delta t \propto \Delta x^2$, the relative errors of the velocity and the stress are expected to behave $\propto \Delta x^2$. The velocity field of the fluid usually is the central observable of interest in LBM simulations, and its second-order convergence has been verified before (e.g., [164]). Apparently, a systematic analysis of the convergence behavior of the deviatoric stress in the LBM has not been conducted until the investigations published in Krüger et al. [165] and Krüger et al. [258]. Since the fluid stress obtained from the LBM plays a central role in this work, it is necessary to better understand its behavior.

In this section, the accuracy of the fluid stress in LBGK simulations is assessed using the example of the decaying Taylor-Green vortex flow (Krüger et al. [258]). The velocity and the stress tensor are known analytically, and the numerical and analytical solutions can be compared locally, i.e., point by point, which allows to define an L2 norm for the relative error.

The decaying Taylor-Green vortex flow is an inhomogeneous, spatially periodic, unsteady solution of the incompressible NSE in the absence of external forces and boundaries,

$$\rho\left(\frac{\partial \boldsymbol{u}}{\partial t} + (\boldsymbol{u}\cdot\nabla)\boldsymbol{u}\right) = -\nabla p + \nabla\cdot\boldsymbol{\sigma} \tag{B.43}$$

with the fluid stress tensor components $\sigma_{\alpha\beta} = \rho\nu(\partial_\alpha u_\beta + \partial_\beta u_\alpha)$. The Taylor-Green vortex flow is a popular case study and has been employed various times in the past to benchmark the LBM (e.g., [128, 149, 267, 268]).

The vortex

For a resting and decaying vortex in 2D, the velocity field reads

$$\boldsymbol{u}(\boldsymbol{x},t) = u_0 \begin{pmatrix} -\sqrt{k_y/k_x}\cos(k_x x)\sin(k_y y) \\ \sqrt{k_x/k_y}\sin(k_x x)\cos(k_y y) \end{pmatrix} e^{-t/t_\mathrm{D}} \tag{B.44}$$

where k_x and k_y are the components of the wave vector $\boldsymbol{k}$. They are computed from the numbers of lattice nodes along the x- and y-axes, $k_x = 2\pi/N_x$ and $k_y = 2\pi/N_y$. The vortex decay time is $t_\mathrm{D} = 1/\left(\nu\left(k_x^2 + k_y^2\right)\right)$. The initial velocity scale is u_0. Using eq. (B.43) and eq. (B.44), it is straightforward to show that the pressure obeys

$$p(\boldsymbol{x},t) = p_0 - \rho\frac{u_0^2}{4}\left(\frac{k_y}{k_x}\cos(2k_x x) + \frac{k_x}{k_y}\cos(2k_y y)\right) e^{-2t/t_\mathrm{D}}. \tag{B.45}$$

The integration constant p_0 can be interpreted as a homogeneous and constant pressure without hydrodynamic significance. It may be set to zero without loss of generality.

Due to its symmetry and tracelessness in incompressible fluids, the deviatoric stress tensor in 2D has only two independent components which are

$$\begin{aligned} \sigma_{xx}(\boldsymbol{x},t) &= 2\rho\nu u_0\sqrt{k_x k_y}\sin(k_x x)\sin(k_y y)e^{-t/t_\mathrm{D}}, \\ \sigma_{xy}(\boldsymbol{x},t) &= \rho\nu u_0\left(\sqrt{k_x^3/k_y} - \sqrt{k_y^3/k_x}\right)\cos(k_x x)\cos(k_y y)e^{-t/t_\mathrm{D}} \end{aligned} \tag{B.46}$$

for the Taylor-Green vortex flow. The remaining components are $\sigma_{yy} = -\sigma_{xx}$ and $\sigma_{yx} = \sigma_{xy}$. It has to be noted that σ_{xy} vanishes if $k_x = k_y$. Additionally, in the current flow configuration, the pressure term balances the advection term, $\rho(\boldsymbol{u}\cdot\nabla)\boldsymbol{u} = -\nabla p$, and the deviatoric stress term balances the partial time derivative, $\rho\frac{\partial \boldsymbol{u}}{\partial t} = \nabla\cdot\boldsymbol{\sigma}$.

In order to intensify the benchmark, a constant velocity $\boldsymbol{u}^\mathrm{c} = u_0^\mathrm{c}(\cos\theta, \sin\theta)^\mathrm{T}$ is added to the flow field in eq. (B.44). As a consequence, the vortex is steadily advected in direction $\boldsymbol{u}^\mathrm{c}$ without changing its shape or decay rate. The inclination angle θ with respect to the x-axis is arbitrary. The translational shift of the vortex at time t is $\boldsymbol{x}^\mathrm{s} = \boldsymbol{u}^\mathrm{c}t$. This way, the influence of a Galilean transformation can be investigated.

Tab. B.2.: Convergence rates of velocity and stress in the Taylor-Green vortex flow. The rates are extracted from a linear fit to the log-log data of the errors ϵ_u, $\epsilon_{\sigma_{xx}}$, and $\epsilon_{\sigma_{xy}}$. Overall second-order convergence is evident.

benchmark series	convergence for		
	$\boldsymbol{u}$	σ_{xx}	σ_{xy}
reference ($u_0^c = 0$, $\tau = 0.8$)	1.99	1.95	2.00
$\theta = 0°$	1.99	1.98	2.01
$\theta = 17.8°$	1.99	1.97	2.01
$\theta = 45°$	2.00	1.97	2.01
$\tau = 0.51$	2.01	2.01	2.01
$\tau = 0.6$	2.00	2.00	1.98
$\tau = 0.8$	1.99	1.97	2.01
$\tau = 1$	2.00	2.02	1.98

Since all hydrodynamic observables are exactly known at each point in space and time and the flow is fully periodic, the Taylor-Green vortex flow can be used to benchmark the LBM in the *absence of any wall effects.*

Initialization

Owing to its non-trivial time dependence, the correct initialization of the flow is of critical importance (section 5.3.1). The LBM populations f_i are initialized at time $t = 0$ by specifying the equilibrium part $f_i^{(0)} = f_i^{\text{eq}}(\rho, \boldsymbol{u})$, eq. (5.6), and the non-equilibrium part, $f_i^{(1)} \approx f_i^{\text{neq}}(\boldsymbol{\sigma})$, eq. (5.20). Higher order populations are neglected. The required velocity $\boldsymbol{u}$ and stress $\boldsymbol{\sigma}$ are taken from eq. (B.44) and eq. (B.46) at $t = 0$. It has to be emphasized that (i) no external forces are included and (ii) the advection term of the NSE plays an important role here. Thus, the quadratic equilibrium, eq. (5.6), has to be taken.

Since the LBGK is a compressible model with the equation of state $p = c_s^2 \rho$, a pressure gradient is equivalent to a density gradient. This has to be taken into account accordingly. Following eq. (B.45), the density is initialized to

$$\rho(\boldsymbol{x}) = \rho_0 \left[1 - \frac{u_0^2}{4c_s^2}\left(\frac{k_y}{k_x}\cos(2k_x x) + \frac{k_x}{k_y}\cos(2k_y y)\right)\right] \tag{B.47}$$

where ρ_0 is the average fluid density.

In order to quantify the deviations between the numerical and the analytic solutions, the global L2 errors of velocity and stress

$$\epsilon_u(t) := \sqrt{\frac{\sum_{\boldsymbol{x}} \left(\boldsymbol{u}_\text{s}^*(\boldsymbol{x}, t) - \boldsymbol{u}_\text{a}^*(\boldsymbol{x}, t)\right)^2}{\sum_{\boldsymbol{x}} \left(\boldsymbol{u}_\text{a}^*(\boldsymbol{x}, t)\right)^2}}, \quad \epsilon_{\sigma_{xx}}(t) := \sqrt{\frac{\sum_{\boldsymbol{x}} \left(\sigma_{xx,\text{s}}(\boldsymbol{x}, t) - \sigma_{xx,\text{a}}(\boldsymbol{x}, t)\right)^2}{\sum_{\boldsymbol{x}} \left(\sigma_{xx,\text{a}}(\boldsymbol{x}, t)\right)^2}} \tag{B.48}$$

are defined where $\boldsymbol{u}_{\text{s,a}}^*(\boldsymbol{x}, t) := \boldsymbol{u}_{\text{s,a}}(\boldsymbol{x}, t) - \boldsymbol{u}^\text{c}$ is the advection-free part of the velocity field (indices 'a' and 's' denote analytic and simulation results, respectively). The subtraction of $\boldsymbol{u}^\text{c}$ has no effect on the enumerator, but it helps to avoid an advection-biased increase of the reference velocity in the denominator which would lead to an artificial decrease of the relative error. The sum goes over the entire lattice (one unit cell of the periodic flow). The error for σ_{xy} is independently computed in the same way. The stress tensor is not changed by the presence of the advection velocity $\boldsymbol{u}^\text{c}$, except for the time-dependent translational shift $\boldsymbol{x}^\text{s}$.

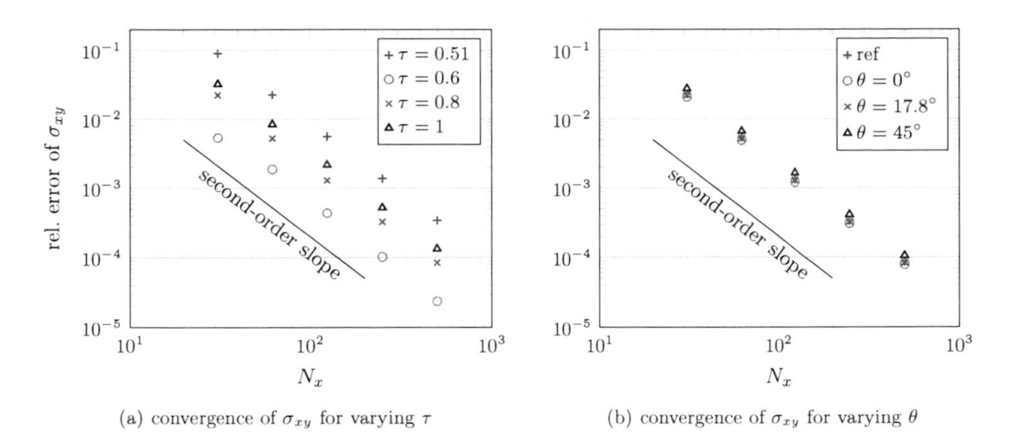

(a) convergence of σ_{xy} for varying τ

(b) convergence of σ_{xy} for varying θ

Fig. B.1.: Convergence of the stress in the Taylor-Green vortex flow. (a) The convergence behavior is shown for four different relaxation parameters ($\tau = 0.51$, 0.6, 0.8, and 1) for the xy-component of the stress tensor. In (b), the convergence behavior for σ_{xy} is presented for three different advection angles ($\theta = 0°$, 17.8°, and 45°) and the reference simulation. The data for σ_{xx} is not shown because it is qualitatively identical.

Simulations and results

For this benchmark test, the D2Q9 lattice (section 5.2) has been employed. All simulations have been performed in series of five simulations each, ranging from a lattice size of $N_x \times N_y = 31 \times 17$ up to 496×272. Between successive simulations, the number of lattice nodes along each axis is multiplied by 2. Within each series, τ is fixed, the time step scales like $\Delta t \propto \Delta x^2$ and the velocity like $u \propto \Delta x$ ('diffusive scaling'). For the coarsest resolution ($N_x \times N_y = 31 \times 17$), the initial velocity scale is $u_0 = \sqrt{0.001}$ in all cases. The impact of the choice of the relaxation parameter τ and the influence of the inclination angle θ on the errors and the convergence rate have been tested.

In one benchmark, four different values of τ (0.51, 0.6, 0.8, and 1) have been used. The remaining parameters are $u_0^c = 0.05$ for the coarsest resolution and $\theta = 17.8°$. In the other benchmark, θ is varied (0°, 17.8°, and 45°) where the remaining parameters are $u_0^c = 0.05$ (coarsest resolution) and $\tau = 0.8$. As a reference, the case without advection has also been tested: $u_0^c = 0$ and $\tau = 0.8$. The average density is $\rho_0 = 1$ in all simulations.

The errors ϵ_u, $\epsilon_{\sigma_{xx}}$, and $\epsilon_{\sigma_{xy}}$ are evaluated at $t = t_D$. The errors for σ_{xy} are shown in fig. B.1, and the extracted convergence rates for the velocity and the stress components are collected in tab. B.2. Without exception, the velocity and the stresses converge with a second-order rate. Neither the presence of the advection velocity $\boldsymbol{u}^c$ along an arbitrary direction nor the non-divisibility of the channel dimensions (N_x, N_y) have a significant effect on the stress errors. The choice of the relaxation time however, alters the magnitude of stress errors even though the convergence rate still remains of second order. The largest errors are found for $\tau = 0.51$, which seems to indicate the vicinity to the numerical instability at τ close to 0.5.

Concluding, apart from the recovered second-order convergence of the velocity, the above findings provide direct evidence for the predicted second-order convergence of the stress tensor in LBGK simulations, as discussed in appx. B.1.3.

B.2. Recovery and initialization of non-equilibrium populations

In the absence of forces, the non-equilibrium populations can be approximated by, cf. eq. (5.20), [126]

$$f_i^{\mathrm{neq}}(\boldsymbol{\sigma}) = -\frac{w_i}{2c_s^4}\frac{1}{1-\frac{1}{2\tau}}\boldsymbol{Q}_i : \boldsymbol{\sigma}. \tag{B.49}$$

With the isotropy relations in eq. (5.8), it is straightforward to verify that the populations in eq. (B.49) yield

$$\sum_i f_i^{\mathrm{neq}} = 0, \quad \sum_i f_i^{\mathrm{neq}}\boldsymbol{c}_i = 0, \quad \sum_i f_i^{\mathrm{neq}}\boldsymbol{c}_i\boldsymbol{c}_i = -\frac{1}{1-\frac{1}{2\tau}}\boldsymbol{\sigma} \tag{B.50}$$

which are the correct results from eq. (5.14), eq. (5.15), and eq. (5.17) in the absence of forces.

Body force densities give rise to correction terms for the velocity, eq. (5.15), and the stress, eq. (5.17). These correction terms are contained in the non-equilibrium populations, cf. appx. B.1.1, which indicates that the non-equilibrium populations in eq. (B.49) have to be modified in order to obtain the correct macroscopic observables.

For the quadratic equilibrium in eq. (5.6), the corrected populations read, cf. eq. (5.21),

$$f_i^{\mathrm{neq}}(\boldsymbol{u}, \boldsymbol{\sigma}, \boldsymbol{f}) = -\frac{w_i}{2c_s^4}\frac{1}{1-\frac{1}{2\tau}}\boldsymbol{Q}_i : \boldsymbol{\sigma} - \frac{w_i \Delta t}{2c_s^2}\boldsymbol{c}_i \cdot \boldsymbol{f} - \frac{w_i \Delta t}{4c_s^4}\boldsymbol{Q}_i : (\boldsymbol{u}\boldsymbol{f} + \boldsymbol{f}\boldsymbol{u}), \tag{B.51}$$

and for the linearized equilibrium in eq. (5.9), cf. eq. (5.22),

$$f_i^{\mathrm{neq}}(\boldsymbol{u}, \boldsymbol{\sigma}, \boldsymbol{f}) = -\frac{w_i}{2c_s^4}\frac{1}{1-\frac{1}{2\tau}}\boldsymbol{Q}_i : \boldsymbol{\sigma} - \frac{w_i \Delta t}{2c_s^2}\boldsymbol{c}_i \cdot \boldsymbol{f}. \tag{B.52}$$

Instead of deriving the correction terms from underlying assumptions [269], it is shown that these terms indeed provide the correct relations between moments of the non-equilibrium populations and the macroscopic observables. Again exploiting the isotropy relations in eq. (5.8), the first three moments of the second term on the right-hand-sides of eq. (B.51) and eq. (B.52) read

$$\frac{\Delta t}{2c_s^2}\sum_i w_i c_{i\alpha} f_\alpha = 0, \quad \frac{\Delta t}{2c_s^2}\sum_i w_i c_{i\alpha} c_{i\beta} f_\alpha = \frac{\Delta t}{2}\delta_{\alpha\beta} f_\alpha, \quad \frac{\Delta t}{2c_s^2}\sum_i w_i c_{i\alpha} c_{i\beta} c_{i\mu} f_\alpha = 0. \tag{B.53}$$

Thus, it can be seen that this term introduces the expected correction term for the forcing in the velocity, eq. (5.15). A similar analysis reveals that the first three moments of the third term on the right-hand-side of eq. (B.51) read

$$\begin{gathered}\frac{\Delta t}{4c_s^4}\sum_i w_i Q_{i\alpha\beta}(u_\alpha f_\beta + f_\alpha u_\beta) = 0, \quad \frac{\Delta t}{4c_s^4}\sum_i w_i Q_{i\alpha\beta} c_{i\mu}(u_\alpha f_\beta + f_\alpha u_\beta) = 0, \\ \frac{\Delta t}{4c_s^4}\sum_i w_i Q_{i\alpha\beta} c_{i\mu} c_{i\nu}(u_\alpha f_\beta + f_\alpha u_\beta) = \frac{\Delta t}{4}(\delta_{\alpha\mu}\delta_{\beta\nu} + \delta_{\alpha\nu}\delta_{\beta\mu})(u_\alpha f_\beta + f_\alpha u_\beta).\end{gathered} \tag{B.54}$$

Here, the correction term for the stress in eq. (5.17) is recovered. Obviously, there is no additional correction term for the linearized equilibrium.

Concluding, eq. (B.51) and eq. (B.52) describe a self-consistent reconstruction for the non-equilibrium populations which can be used for initialization of LBM simulations in the presence of a body force if the velocity and the stress are explicitly known.

C. Derivation of the membrane forces

The membrane forces due to strain, bending, surface, and volume constraints are computed from the corresponding energy terms with the principle of virtual work. This procedure has been employed by various researchers (e.g., [186, 192, 195, 196, 243, 270, 271]).

Given a discretized membrane energy term in the form $E(\{\boldsymbol{x}_i\})$, the discretized force acting on node i against the deformation is computed from

$$\boldsymbol{F}_i = -\frac{\partial E(\{\boldsymbol{x}_i\})}{\partial \boldsymbol{x}_i} \tag{C.1}$$

with all other positions $\{\boldsymbol{x}_j\}$, $j \neq i$, fixed. The above approach bases on the assumption that accelerations are negligible, i.e., the membrane is in mechanical equilibrium at all times. In the present thesis, inertia effects are commonly disregarded.

If the energy $E(\{\boldsymbol{x}_i\})$ is differentiable with respect to the positions $\{\boldsymbol{x}_i\}$, the derivatives may be precomputed and an expensive numerical differentiation can be avoided. For the implementation of the presented membrane model, all necessary derivatives have been precomputed analytically. It has turned out that this approach—although more demanding on the developer's side—is numerically much more efficient than finding the forces numerically via 'shaking' of the energy about its current value.

The derivations of the forces are presented in the following sections: strain and dilation in appx. C.1, bending in appx. C.2, surface in appx. C.3, and volume in appx. C.4.

C.1. Derivation of the strain force

C.1.1. Displacement gradient tensor

As explained in section 7.1, the faces of an undeformed and a deformed face element may be rotated in such a way that the problem is 2D, and any z-coordinates can be dropped, cf. fig. C.1. Both faces shall be aligned along one side and share one node. This is no restriction since the strain energy of the face is invariant under rotations and translations. Any properties of the undeformed face are denoted by a superscript (0).

From fig. C.1, one can find the coordinates of the nodes in the undeformed face,

$$\boldsymbol{x}_1^{(0)} = \begin{pmatrix} 0 \\ 0 \end{pmatrix}, \quad \boldsymbol{x}_2^{(0)} = \begin{pmatrix} l'^{(0)} \\ 0 \end{pmatrix}, \quad \boldsymbol{x}_3^{(0)} = \begin{pmatrix} l^{(0)} \cos\varphi^{(0)} \\ l^{(0)} \sin\varphi^{(0)} \end{pmatrix}, \tag{C.2}$$

and in the deformed face,

$$\boldsymbol{x}_1 = \begin{pmatrix} 0 \\ 0 \end{pmatrix}, \quad \boldsymbol{x}_2 = \begin{pmatrix} l' \\ 0 \end{pmatrix}, \quad \boldsymbol{x}_3 = \begin{pmatrix} l \cos\varphi \\ l \sin\varphi \end{pmatrix}. \tag{C.3}$$

The three displacement vectors $\boldsymbol{V}_{1,2,3} = \boldsymbol{x}_{1,2,3} - \boldsymbol{x}_{1,2,3}^{(0)}$, also shown in fig. C.1, read

$$\boldsymbol{V}_1 = \begin{pmatrix} 0 \\ 0 \end{pmatrix}, \quad \boldsymbol{V}_2 = \begin{pmatrix} l' - l'^{(0)} \\ 0 \end{pmatrix}, \quad \boldsymbol{V}_3 = \begin{pmatrix} l\cos\varphi - l^{(0)}\cos\varphi^{(0)} \\ l\sin\varphi - l^{(0)}\sin\varphi^{(0)} \end{pmatrix}. \tag{C.4}$$

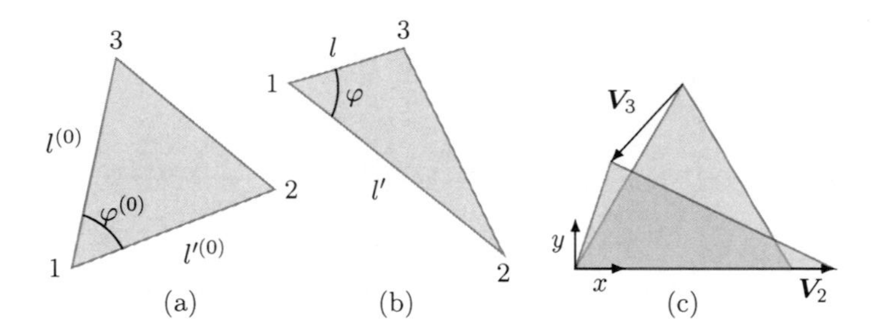

Fig. C.1.: Deformation of a membrane face element. Each face is made up of three nodes (1, 2, 3) which define the edges l (between nodes 1 and 3) and l' (between nodes 1 and 2) and the angle φ between these two edges. (a) The equilibrium face (defined by $l^{(0)}$, $l'^{(0)}$, and $\varphi^{(0)}$), (b) its deformed shape (accordingly defined by l, l', and φ), and (c) both transformed to the same xy-plane are shown. The displacement vector $\boldsymbol{V}_1$ is identically zero, and the other two are shown in (c). The deformation state (λ_1, λ_2) of the face is then uniquely defined.

In order to transform the deformation state of the nodes to a deformation state of the face, some tools of the finite element method (FEM) are employed. Three linear shape functions $N_{1,2,3}(x,y)$ are used to interpolate the displacement at any position (x, y) in the face, e.g., [272],

$$\boldsymbol{V}(x,y) = N_1(x,y)\boldsymbol{V}_1 + N_2(x,y)\boldsymbol{V}_2 + N_3(x,y)\boldsymbol{V}_3, \tag{C.5}$$

where

$$N_{1,2,3}(x,y) = a_{1,2,3}x + b_{1,2,3}y + c_{1,2,3}. \tag{C.6}$$

The constants $a_{1,2,3}$, $b_{1,2,3}$, and $c_{1,2,3}$ only depend on the undeformed face geometry and do not change in time. They can be found by letting $N_j(\boldsymbol{x}_k) = \delta_{jk}$. The solution is

$$\begin{aligned}
a_1 &= \frac{y_2^{(0)} - y_3^{(0)}}{2A^{(0)}} = -\frac{l^{(0)}\sin\varphi^{(0)}}{2A^{(0)}}, & b_1 &= \frac{x_3^{(0)} - x_2^{(0)}}{2A^{(0)}} = \frac{l^{(0)}\cos\varphi^{(0)} - l'^{(0)}}{2A^{(0)}}, \\
a_2 &= \frac{y_3^{(0)} - y_1^{(0)}}{2A^{(0)}} = \frac{l^{(0)}\sin\varphi^{(0)}}{2A^{(0)}}, & b_2 &= \frac{x_1^{(0)} - x_3^{(0)}}{2A^{(0)}} = -\frac{l^{(0)}\cos\varphi^{(0)}}{2A^{(0)}}, \\
a_3 &= \frac{y_1^{(0)} - y_2^{(0)}}{2A^{(0)}} = 0, & b_3 &= \frac{x_2^{(0)} - x_1^{(0)}}{2A^{(0)}} = \frac{l'^{(0)}}{2A^{(0)}}.
\end{aligned} \tag{C.7}$$

Here, $A^{(0)} = \frac{1}{2}l^{(0)}l'^{(0)}\cos\varphi^{(0)}$ is the area of the undeformed face which is required for the normalization of the shape functions. Since only the derivatives of the shape functions will be required for the further calculations, the three constants $c_{1,2,3}$ are not shown here. The idea behind the introduction of the shape functions is that the deformation is assumed to be linear across the entire face, which means that the deformation gradient is constant. It is also possible to consider higher orders in the deformation gradient [191, 194] increasing both the accuracy and the computational costs.

The surface deformation gradient tensor is defined as [189]

$$\boldsymbol{D} = \begin{pmatrix} D_{xx} & D_{xy} \\ D_{yx} & D_{yy} \end{pmatrix} := \begin{pmatrix} 1 & 0 \\ 0 & 1 \end{pmatrix} + \begin{pmatrix} \partial_x V_x & \partial_y V_x \\ \partial_x V_y & \partial_y V_y \end{pmatrix}, \tag{C.8}$$

and the derivatives evaluate to

$$\begin{aligned}
D_{xx} &= 1 + & a_1 V_{1x} + a_2 V_{2x} + a_3 V_{3x} &= \frac{l'}{l'^{(0)}}, \\
D_{xy} &= & b_1 V_{1x} + b_2 V_{2x} + b_3 V_{3x} &= \frac{1}{\sin\varphi^{(0)}}\left(\frac{l}{l^{(0)}}\cos\varphi - \frac{l'}{l'^{(0)}}\cos\varphi^{(0)}\right), \\
D_{yx} &= & a_1 V_{1y} + a_2 V_{2y} + a_3 V_{3y} &= 0, \\
D_{yy} &= 1 + & b_1 V_{1y} + b_2 V_{2y} + b_3 V_{3y} &= \frac{l}{l^{(0)}}\frac{\sin\varphi}{\sin\varphi^{(0)}}.
\end{aligned} \tag{C.9}$$

Once the undeformed and the deformed positions of the nodes in a face are known, the linearly approximated deformation gradient tensor $\boldsymbol{D}$ can be computed from eq. (C.9) and the strain invariants I_1 and I_2 and thus the face energy ϵ^{S} are uniquely determined as described in section 7.1.

C.1.2. Strain force

The strain force acting on a node i at position $\boldsymbol{x}_i$ can be computed from the strain energy $E_{\mathrm{S}} = \sum_j A_j^{(0)} \epsilon_j^{\mathrm{S}}$ via the principle of virtual work,

$$\boldsymbol{F}_i^{\mathrm{S}} = -\frac{\partial E_{\mathrm{S}}(\{\boldsymbol{x}_i\})}{\partial \boldsymbol{x}_i}. \tag{C.10}$$

The strain energy E_{S} is uniquely determined when all node positions $\{\boldsymbol{x}_i\}$ of the mesh are known. In the following, i is a node index and j is a face index. Obviously, the energy $A_j^{(0)} \epsilon_j^{\mathrm{S}}$ of face j depends only on the positions of the three nodes which belong to this particular face, cf. appx. C.1.1. This makes the problem local and simplifies the numerical implementation. Due to symmetry arguments, cf. section 7.1, the strain energy density is a function of the strain invariants I_1 and I_2 only. The actual form $\epsilon^{\mathrm{S}} = \epsilon^{\mathrm{S}}(I_1, I_2)$ of this function bases on an underlying physical model and is not predicted by the theory of elasticity or the employed FEM. One possible model—which is taken for most simulations in this thesis—is Skalak's law, eq. (7.2).

The strain invariants derive from the squared displacement gradient tensor [189],

$$\begin{aligned}
I_1 &= \operatorname{tr}\boldsymbol{G} - 2, \\
I_2 &= \det\boldsymbol{G} - 1,
\end{aligned} \tag{C.11}$$

where the symmetric tensor $\boldsymbol{G}$ obeys

$$\boldsymbol{G} = \boldsymbol{D}^{\mathrm{T}}\boldsymbol{D} = \begin{pmatrix} D_{xx}^2 + D_{yx}^2 & D_{xx}D_{xy} + D_{yx}D_{yy} \\ D_{xx}D_{xy} + D_{yy}D_{yx} & D_{xy}^2 + D_{yy}^2 \end{pmatrix}. \tag{C.12}$$

Within the FEM, the linearized displacement gradient tensor $\boldsymbol{D}$ is given in eq. (C.9). According to this equation, $\boldsymbol{D}$ is a function of the three node displacements $\boldsymbol{V}_{1,2,3}$.

The strain force can only act in the plane which is defined by the face. For this reason, the entire problem may be formulated in 2D, and the strain force in the rotated coordinate system, cf. fig. C.1, is

$$\boldsymbol{F}_i^{\mathrm{S,rot}} = -\frac{\partial E_{\mathrm{S}}(\{I_{1j}, I_{2j}\})}{\partial \boldsymbol{V}_i} \tag{C.13}$$

with $I_{1j} = I_{1j}(\boldsymbol{D}_j)$, $I_{2j} = I_{2j}(\boldsymbol{D}_j)$, and $\boldsymbol{D}_j = \boldsymbol{D}_j(\{\boldsymbol{V}_i\})$. It is important to mention again that, by definition, the displacement vectors $\boldsymbol{V}_{1,2,3}$ in the rotated face do not have a z-component. The z-component of $\boldsymbol{F}_i^{\mathrm{S,rot}}$, therefore, is identically zero.

It is straightforward but tedious to compute the strain force $\boldsymbol{F}_i^{\mathrm{S,rot}}$ with nested chain rules from the energy density ϵ_j^{S}. This procedure will be shown in detail in the following. The first step is the computation of $\partial\epsilon^{\mathrm{S}}/\partial I_1$ and $\partial\epsilon^{\mathrm{S}}/\partial I_2$. These derivatives depend on the constitutive model. For Skalak's law, they read

$$\frac{\partial \epsilon^{\mathrm{S}}}{\partial I_1} = \frac{\kappa_{\mathrm{S}}}{6}(I_1 + 1), \qquad \frac{\partial \epsilon^{\mathrm{S}}}{\partial I_2} = -\frac{\kappa_{\mathrm{S}}}{6} + \frac{\kappa_\alpha}{6} I_2. \tag{C.14}$$

From eq. (C.11) follow the derivatives

$$\begin{aligned}
&\frac{\partial I_1}{\partial G_{xx}} = 1, &&\frac{\partial I_1}{\partial G_{xy}} = 0, &&\frac{\partial I_1}{\partial G_{yy}} = 1,\\
&\frac{\partial I_2}{\partial G_{xx}} = G_{yy}, &&\frac{\partial I_2}{\partial G_{xy}} = -2G_{xy}, &&\frac{\partial I_2}{\partial G_{yy}} = G_{xx}.
\end{aligned} \tag{C.15}$$

Exploiting the relations in eq. (C.12) and eq. (C.9), the derivatives of G_{xx} with respect to the displacements $\boldsymbol{V}$ are

$$\begin{aligned}
\frac{\partial G_{xx}}{\partial V_{1x}} &= 2D_{xx}\frac{\partial D_{xx}}{\partial V_{1x}} + 2D_{yx}\frac{\partial D_{yx}}{\partial V_{1x}} = 2a_1 D_{xx}, & \frac{\partial G_{xx}}{\partial V_{1y}} &= 2D_{xx}\frac{\partial D_{xx}}{\partial V_{1y}} + 2D_{yx}\frac{\partial D_{yx}}{\partial V_{1y}} = 0,\\
\frac{\partial G_{xx}}{\partial V_{2x}} &= 2D_{xx}\frac{\partial D_{xx}}{\partial V_{2x}} + 2D_{yx}\frac{\partial D_{yx}}{\partial V_{2x}} = 2a_2 D_{xx}, & \frac{\partial G_{xx}}{\partial V_{2y}} &= 2D_{xx}\frac{\partial D_{xx}}{\partial V_{2y}} + 2D_{yx}\frac{\partial D_{yx}}{\partial V_{2y}} = 0,\\
\frac{\partial G_{xx}}{\partial V_{3x}} &= 2D_{xx}\frac{\partial D_{xx}}{\partial V_{3x}} + 2D_{yx}\frac{\partial D_{yx}}{\partial V_{3x}} = 2a_3 D_{xx}, & \frac{\partial G_{xx}}{\partial V_{3y}} &= 2D_{xx}\frac{\partial D_{xx}}{\partial V_{3y}} + 2D_{yx}\frac{\partial D_{yx}}{\partial V_{3y}} = 0.
\end{aligned} \tag{C.16}$$

For the derivatives of G_{xy} one finds

$$\begin{aligned}
\frac{\partial G_{xy}}{\partial V_{1x}} &= D_{xy}\frac{\partial D_{xx}}{\partial V_{1x}} + D_{xx}\frac{\partial D_{xy}}{\partial V_{1x}} + D_{yx}\frac{\partial D_{yy}}{\partial V_{1x}} + D_{yy}\frac{\partial D_{yx}}{\partial V_{1x}} = a_1 D_{xy} + b_1 D_{xx},\\
\frac{\partial G_{xy}}{\partial V_{1y}} &= D_{xy}\frac{\partial D_{xx}}{\partial V_{1y}} + D_{xx}\frac{\partial D_{xy}}{\partial V_{1y}} + D_{yx}\frac{\partial D_{yy}}{\partial V_{1y}} + D_{yy}\frac{\partial D_{yx}}{\partial V_{1y}} = a_1 D_{yy},\\
\frac{\partial G_{xy}}{\partial V_{2x}} &= D_{xy}\frac{\partial D_{xx}}{\partial V_{2x}} + D_{xx}\frac{\partial D_{xy}}{\partial V_{2x}} + D_{yx}\frac{\partial D_{yy}}{\partial V_{2x}} + D_{yy}\frac{\partial D_{yx}}{\partial V_{2x}} = a_2 D_{xy} + b_2 D_{xx},\\
\frac{\partial G_{xy}}{\partial V_{2y}} &= D_{xy}\frac{\partial D_{xx}}{\partial V_{2y}} + D_{xx}\frac{\partial D_{xy}}{\partial V_{2y}} + D_{yx}\frac{\partial D_{yy}}{\partial V_{2y}} + D_{yy}\frac{\partial D_{yx}}{\partial V_{2y}} = a_2 D_{yy},\\
\frac{\partial G_{xy}}{\partial V_{3x}} &= D_{xy}\frac{\partial D_{xx}}{\partial V_{3x}} + D_{xx}\frac{\partial D_{xy}}{\partial V_{3x}} + D_{yx}\frac{\partial D_{yy}}{\partial V_{3x}} + D_{yy}\frac{\partial D_{yx}}{\partial V_{3x}} = a_3 D_{xy} + b_3 D_{xx},\\
\frac{\partial G_{xy}}{\partial V_{3y}} &= D_{xy}\frac{\partial D_{xx}}{\partial V_{3y}} + D_{xx}\frac{\partial D_{xy}}{\partial V_{3y}} + D_{yx}\frac{\partial D_{yy}}{\partial V_{3y}} + D_{yy}\frac{\partial D_{yx}}{\partial V_{3y}} = a_3 D_{yy}.
\end{aligned} \tag{C.17}$$

Finally, the derivatives of G_{yy} are

$$\begin{aligned}
\frac{\partial G_{yy}}{\partial V_{1x}} &= 2D_{xy}\frac{\partial D_{xy}}{\partial V_{1x}} + 2D_{yy}\frac{\partial D_{yy}}{\partial V_{1x}} = 2b_1 D_{xy}, & \frac{\partial G_{yy}}{\partial V_{1y}} &= 2D_{xy}\frac{\partial D_{xy}}{\partial V_{1y}} + 2D_{yy}\frac{\partial D_{yy}}{\partial V_{1y}} = 2b_1 D_{yy},\\
\frac{\partial G_{yy}}{\partial V_{2x}} &= 2D_{xy}\frac{\partial D_{xy}}{\partial V_{2x}} + 2D_{yy}\frac{\partial D_{yy}}{\partial V_{2x}} = 2b_2 D_{xy}, & \frac{\partial G_{yy}}{\partial V_{2y}} &= 2D_{xy}\frac{\partial D_{xy}}{\partial V_{2y}} + 2D_{yy}\frac{\partial D_{yy}}{\partial V_{2y}} = 2b_2 D_{yy},\\
\frac{\partial G_{yy}}{\partial V_{3x}} &= 2D_{xy}\frac{\partial D_{xy}}{\partial V_{3x}} + 2D_{yy}\frac{\partial D_{yy}}{\partial V_{3x}} = 2b_3 D_{xy}, & \frac{\partial G_{yy}}{\partial V_{3y}} &= 2D_{xy}\frac{\partial D_{xy}}{\partial V_{3y}} + 2D_{yy}\frac{\partial D_{yy}}{\partial V_{3y}} = 2b_3 D_{yy}.
\end{aligned} \tag{C.18}$$

It is straightforward to put the results together in order to obtain the explicit equations for the forces acting on the three nodes in a given face. For code efficiency, it is advantageous to drop any terms which are identically zero, e.g., $\partial G_{xx}/\partial V_{1y}$. Finally, the forces in the common xy-plane are recovered. The efficiency can further be boosted by taking into account that the sum of the forces in a face is zero. The reason is the symmetry of the strain energy due to momentum conservation. It turns out that the computation of the force $\boldsymbol{F}_2^{\text{S,rot}}$ is most expensive, hence it is rewarding to first compute $\boldsymbol{F}_1^{\text{S,rot}}$ and $\boldsymbol{F}_3^{\text{S,rot}}$ and then $\boldsymbol{F}_2^{\text{S,rot}} = -\boldsymbol{F}_1^{\text{S,rot}} - \boldsymbol{F}_3^{\text{S,rot}}$. In order to complete the strain force evaluation, the forces acting on the nodes in a face have to be rotated back to the correct orientation in 3D.

C.2. Derivation of the bending force

The bending energy model employed in this thesis as given in eq. (7.8) is

$$E_{\text{B}} = \frac{\tilde{\kappa}_{\text{B}}}{2} \sum_{\langle i,j \rangle} \left(\theta_{ij} - \theta_{ij}^{(0)} \right)^2 . \tag{C.19}$$

Since the energy is additive, only one pair of faces with energy E_{ij}^{B} is taken into account here. The total bending force acting on a membrane node is recovered by summing over all pairs of neighboring faces containing this node. For a node k being a member of the pair $\langle i,j \rangle$, the force due to the deformation is

$$\boldsymbol{F}_k^{\text{B}} = -\frac{E_{ij}^{\text{B}}}{\partial \boldsymbol{x}_k} = -\tilde{\kappa}_{\text{B}} \left(\theta_{ij} - \theta_{ij}^{(0)} \right) \frac{\partial \theta_{ij}}{\partial \boldsymbol{x}_k}. \tag{C.20}$$

The derivative can be evaluated with the chain rule,

$$\frac{\partial \theta_{ij}}{\partial \boldsymbol{x}_k} = \frac{\partial \arccos(\boldsymbol{n}_i \cdot \boldsymbol{n}_j)}{\partial \boldsymbol{x}_k} = -\frac{1}{\sqrt{1 - (\boldsymbol{n}_i \cdot \boldsymbol{n}_j)^2}} \frac{\partial (\boldsymbol{n}_i \cdot \boldsymbol{n}_j)}{\partial \boldsymbol{x}_k}. \tag{C.21}$$

It is sufficient to compute only the force for one node in the pair of faces (node 1 is considered, cf. fig. C.2). The corresponding forces acting on the other three nodes in the pair of faces can simply be found by changing the indices.

In order to evaluate the derivative of the unit normal vector $\boldsymbol{n}_i$, the normal vector

$$\boldsymbol{N}_i = (\boldsymbol{x}_{i1} - \boldsymbol{x}_{i3}) \times (\boldsymbol{x}_{i2} - \boldsymbol{x}_{i3}) = 2A_i \boldsymbol{n}_i, \tag{C.22}$$

is considered first, cf. fig. C.2. Here, $\boldsymbol{x}_{i1}$, $\boldsymbol{x}_{i2}$, and $\boldsymbol{x}_{i3}$ are the position vectors of the nodes belonging to face i. They are sorted in such a way that the normal vectors $\boldsymbol{n}_i$ and $\boldsymbol{N}_i$ point outwards. From this, in component notation it follows that

$$\frac{\partial n_{i\beta}}{\partial x_{i1\alpha}} = \frac{\partial}{\partial x_{i1\alpha}} \frac{N_{i\beta}}{2A_i} = \frac{1}{2A_i} \frac{\partial N_{i\beta}}{\partial x_{i1\alpha}} - \frac{N_{i\beta}}{2A_i^2} \frac{\partial A_i}{\partial x_{i1\alpha}} \tag{C.23}$$

and similarly for the vectors $\boldsymbol{x}_{i2}$ and $\boldsymbol{x}_{i3}$. Since $A_i = \sqrt{N_{i\beta} N_{i\beta}}/2$, one can write

$$\frac{\partial A_i}{\partial x_{i1\alpha}} = \frac{N_{i\beta}}{2\sqrt{N_{i\gamma} N_{i\gamma}}} \frac{\partial N_{i\beta}}{\partial x_{i1\alpha}} = \frac{n_{i\beta}}{2} \frac{\partial N_{i\beta}}{\partial x_{i1\alpha}}. \tag{C.24}$$

It remains the evaluation of the derivative of $\boldsymbol{N}$,

$$\frac{\partial N_{i\beta}}{\partial x_{i1\alpha}} = \frac{\partial}{\partial x_{i1\alpha}} \left[\epsilon_{\beta\rho\sigma} (x_{i1\rho} - x_{i3\rho})(x_{i2\sigma} - x_{i3\sigma}) \right] = \epsilon_{\beta\rho\sigma} \delta_{\alpha\rho} (x_{i2\sigma} - x_{i3\sigma}) = \epsilon_{\beta\alpha\sigma} (x_{i2\sigma} - x_{i3\sigma}). \tag{C.25}$$

One can rewrite eq. (C.23) with the help of eq. (C.24),

$$\frac{\partial n_{i\beta}}{\partial x_{i1\alpha}} = \frac{1}{2A_i} \left(\frac{\partial N_{i\beta}}{\partial x_{i1\alpha}} - n_{i\beta} n_{i\gamma} \frac{\partial N_{i\gamma}}{\partial x_{i1\alpha}} \right). \tag{C.26}$$

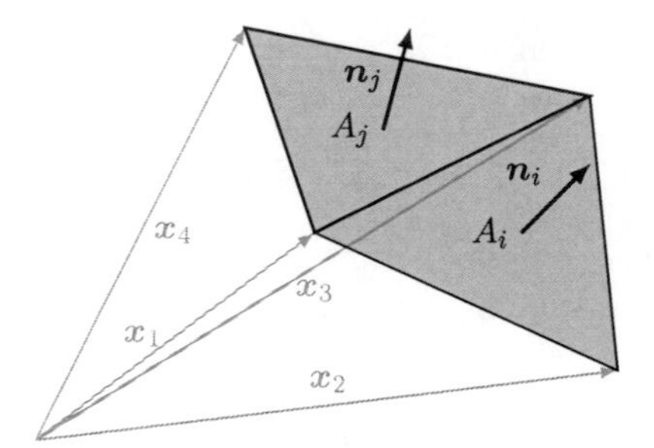

Fig. C.2.: Membrane face and node convention for the derivation of the bending force. The pair $\langle i, j \rangle$ of faces is defined by four nodes with coordinates $\boldsymbol{x}_1, \ldots, \boldsymbol{x}_4$. Nodes 1 and 3 belong to both faces, nodes 2 and 4 only to one. The normal vectors $\boldsymbol{n}_i$ and $\boldsymbol{n}_j$ are defined in such a way that they point outwards.

It directly follows from eq. (C.25) that

$$\begin{aligned} n_{j\beta}\frac{\partial n_{i\beta}}{\partial x_{i1\alpha}} &= \frac{1}{2A_i}\left(\epsilon_{\beta\alpha\sigma}(x_{i2\sigma} - x_{i3\sigma})n_{j\beta} - \epsilon_{\gamma\alpha\sigma}(x_{i2\sigma} - x_{i3\sigma})n_{j\beta}n_{i\beta}n_{i\gamma}\right) \\ &= \frac{1}{2A_i}\epsilon_{\beta\alpha\sigma}(x_{i2\sigma} - x_{i3\sigma})\left(n_{j\beta} - n_{i\gamma}n_{j\gamma}n_{i\beta}\right) \end{aligned} \tag{C.27}$$

and finally

$$\frac{\partial(\boldsymbol{n}_i \cdot \boldsymbol{n}_j)}{\partial \boldsymbol{x}_1} = \frac{1}{2A_i}(\boldsymbol{x}_2 - \boldsymbol{x}_3) \times (\boldsymbol{n}_j - (\boldsymbol{n}_i \cdot \boldsymbol{n}_j)\boldsymbol{n}_i) + \frac{1}{2A_j}(\boldsymbol{x}_3 - \boldsymbol{x}_4) \times (\boldsymbol{n}_i - (\boldsymbol{n}_i \cdot \boldsymbol{n}_j)\boldsymbol{n}_j)\,. \tag{C.28}$$

with the conventions from fig. C.2. The remaining derivatives are

$$\frac{\partial(\boldsymbol{n}_i \cdot \boldsymbol{n}_j)}{\partial \boldsymbol{x}_2} = \frac{1}{2A_i}(\boldsymbol{x}_3 - \boldsymbol{x}_1) \times (\boldsymbol{n}_j - (\boldsymbol{n}_i \cdot \boldsymbol{n}_j)\boldsymbol{n}_i)\,, \tag{C.29}$$

$$\frac{\partial(\boldsymbol{n}_i \cdot \boldsymbol{n}_j)}{\partial \boldsymbol{x}_3} = \frac{1}{2A_i}(\boldsymbol{x}_1 - \boldsymbol{x}_2) \times (\boldsymbol{n}_j - (\boldsymbol{n}_i \cdot \boldsymbol{n}_j)\boldsymbol{n}_i) + \frac{1}{2A_j}(\boldsymbol{x}_4 - \boldsymbol{x}_1) \times (\boldsymbol{n}_i - (\boldsymbol{n}_i \cdot \boldsymbol{n}_j)\boldsymbol{n}_j)\,, \tag{C.30}$$

$$\frac{\partial(\boldsymbol{n}_i \cdot \boldsymbol{n}_j)}{\partial \boldsymbol{x}_4} = \frac{1}{2A_j}(\boldsymbol{x}_1 - \boldsymbol{x}_3) \times (\boldsymbol{n}_i - (\boldsymbol{n}_i \cdot \boldsymbol{n}_j)\boldsymbol{n}_j)\,. \tag{C.31}$$

Nodes 1 and 3 are located in both faces whereas nodes 2 and 4 are member of one face only. As a consequence, $\boldsymbol{F}_1^{\mathrm{B}}$ and $\boldsymbol{F}_3^{\mathrm{B}}$ have two contributions, and $\boldsymbol{F}_2^{\mathrm{B}}$ and $\boldsymbol{F}_4^{\mathrm{B}}$ have only one.

The final expressions for the forces can be simplified owing to the identity

$$|\boldsymbol{n}_i - (\boldsymbol{n}_i \cdot \boldsymbol{n}_j)\boldsymbol{n}_j| = |\boldsymbol{n}_j - (\boldsymbol{n}_i \cdot \boldsymbol{n}_j)\boldsymbol{n}_i| = \sqrt{1 - (\boldsymbol{n}_i \cdot \boldsymbol{n}_j)^2}. \tag{C.32}$$

Thus, the denominator in eq. (C.21) can be canceled.

C.3. Derivation of the surface force

The surface force

$$\boldsymbol{F}_i^{\mathrm{A}} = -\frac{\partial E_{\mathrm{A}}(\{\boldsymbol{x}_i\})}{\partial \boldsymbol{x}_i} \tag{C.33}$$

acting on node i with surface energy (cf. section 7.3)

$$E_{\mathrm{A}} = \frac{\kappa_{\mathrm{A}}}{2}\frac{\left(A - A^{(0)}\right)^2}{A^{(0)}} \tag{C.34}$$

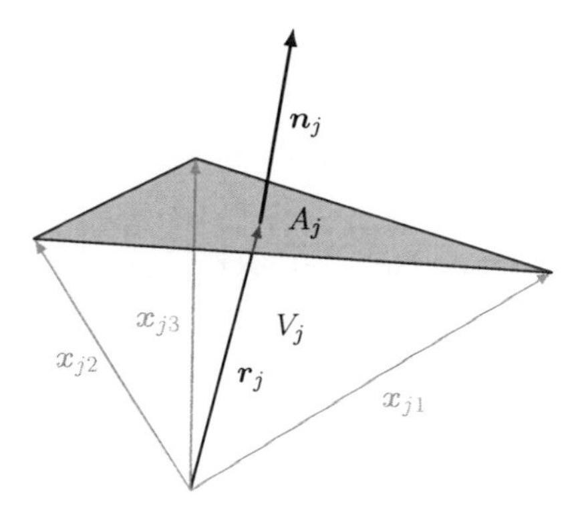

Fig. C.3.: Volume and surface contributions of a single membrane face. The volume V_i of face i can be computed from its area A_j, its unit normal $\boldsymbol{n}_j$, and the center vector $\boldsymbol{r}_j$ or directly from the three node vectors $\boldsymbol{x}_{j1}$, $\boldsymbol{x}_{j2}$, and $\boldsymbol{x}_{j3}$, cf. eq. (C.44).

shall be computed. Thus, the functional dependence $A(\{\boldsymbol{x}_i\})$ must be known and differentiated. The total surface of an arbitrary, closed surface mesh consisting of flat triangular face elements is

$$A = \sum_j A_j. \tag{C.35}$$

The sum runs over all faces j with area

$$A_j = \frac{1}{2}\sqrt{\boldsymbol{N}_j \cdot \boldsymbol{N}_j}. \tag{C.36}$$

Here, $\boldsymbol{N}_j$ is the face normal vector of face j,

$$\boldsymbol{N}_j = \left[(\boldsymbol{x}_{j1} - \boldsymbol{x}_{j2}) \times (\boldsymbol{x}_{j3} - \boldsymbol{x}_{j2})\right], \tag{C.37}$$

and the vectors $\boldsymbol{x}_{j1}$, $\boldsymbol{x}_{j2}$, and $\boldsymbol{x}_{j3}$ are defined in fig. C.3. The unit normal vector is

$$\boldsymbol{n}_j = \frac{\boldsymbol{N}_j}{\sqrt{\boldsymbol{N}_j \cdot \boldsymbol{N}_j}}. \tag{C.38}$$

The surface force acting on node i is

$$\boldsymbol{F}_i^{\mathrm{A}} = -\frac{\partial E_{\mathrm{A}}(\{\boldsymbol{x}_i\})}{\partial \boldsymbol{x}_i} = -\kappa_{\mathrm{A}} \frac{A - A^{(0)}}{A^{(0)}} \frac{\partial A(\{\boldsymbol{x}_i\})}{\partial \boldsymbol{x}_i}. \tag{C.39}$$

A lengthy but straightforward computation reveals that

$$\frac{\partial A_j}{\partial \boldsymbol{x}_{j1}} = \frac{1}{2}\boldsymbol{n}_j \times (\boldsymbol{x}_{j3} - \boldsymbol{x}_{j2}). \tag{C.40}$$

Since eq. (C.37) remains valid under cyclic permutation of the three vectors $\boldsymbol{x}_{j1}$, $\boldsymbol{x}_{j2}$, and $\boldsymbol{x}_{j3}$, analog relations are obtained for the derivatives $\partial A_j / \partial \boldsymbol{x}_{j2}$ and $\partial A_j / \partial \boldsymbol{x}_{j3}$. The total surface force acting on node i then is the sum of contributions of all faces j which node i is a member of.

C.4. Derivation of the volume force

The volume force

$$\boldsymbol{F}_i^{\mathrm{V}} = -\frac{\partial E_{\mathrm{V}}(\{\boldsymbol{x}_i\})}{\partial \boldsymbol{x}_i} \tag{C.41}$$

acting on node i with volume energy (cf. section 7.4)

$$E_{\mathrm{V}} = \frac{\kappa_{\mathrm{V}}}{2} \frac{\left(V - V^{(0)}\right)^2}{V^{(0)}} \tag{C.42}$$

shall be computed. In order to do so, the functional dependence $V(\{\boldsymbol{x}_i\})$ must be known and differentiated. The volume of an arbitrary, closed surface mesh consisting of flat triangular face elements is

$$V = \sum_j V_j. \tag{C.43}$$

The sum runs over all faces (area A_j, unit normal vector $\boldsymbol{n}_j$), and

$$V_j = \frac{1}{3} A_j \boldsymbol{n}_j \cdot \boldsymbol{r}_j = \frac{1}{6} (\boldsymbol{x}_{j3} \times \boldsymbol{x}_{j2}) \cdot \boldsymbol{x}_{j1} \tag{C.44}$$

is the volume contribution assigned to face j[1]. The vector $\boldsymbol{r}_j$ points from the centroid of the mesh to the centroid of face j, and the three vectors $\boldsymbol{x}_{j1}$, $\boldsymbol{x}_{j2}$, and $\boldsymbol{x}_{j3}$ are defined in fig. C.3. The volume force acting on node i is

$$\boldsymbol{F}_i^{\mathrm{V}} = -\frac{\partial E_{\mathrm{V}}(\{\boldsymbol{x}_i\})}{\partial \boldsymbol{x}_i} = -\kappa_{\mathrm{V}} \frac{V - V^{(0)}}{V^{(0)}} \frac{\partial V(\{\boldsymbol{x}_i\})}{\partial \boldsymbol{x}_i}. \tag{C.45}$$

A simple computation reveals that

$$\frac{\partial V_j}{\partial \boldsymbol{x}_{j1}} = \frac{1}{6} (\boldsymbol{x}_{j3} \times \boldsymbol{x}_{j2}). \tag{C.46}$$

The triple product, $(\boldsymbol{x}_{j3} \times \boldsymbol{x}_{j2}) \cdot \boldsymbol{x}_{j1}$, is invariant under cyclic permutations of the three contained vectors. Consequently, similar results are obtained for the derivatives $\partial V_j / \partial \boldsymbol{x}_{j2}$ and $\partial V_j / \partial \boldsymbol{x}_{j3}$. The total volume force acting on node i then is the sum of contributions of all faces j which node i is a member of.

[1] The volume contribution for individual faces can be negative for concave meshes when $\boldsymbol{r}_j \cdot \boldsymbol{n}_j < 0$. This does not limit the general validity of the volume force algorithm.

Bibliography

[1] J.J. Stickel and R.L. Powell. Fluid mechanics and rheology of dense suspensions. *Annu. Rev. Fluid Mech.*, 37:129–149, 2005. Cited on pages 5, 6, 7, 8, 9, 95, and 107.

[2] L. D. Landau, E. M. Lifshits, and W. Weller. *Hydrodynamik*. Akademie-Verlag, Berlin, 1966. Cited on pages 5, 73, and 74.

[3] R.G. Larson. *The Structure and Rheology of Complex Fluids*. Oxford University Press, 1999. ISBN 019512197X. Cited on pages 6, 7, 8, 9, and 107.

[4] B.D. Todd, D.J. Evans, and P.J. Daivis. Pressure tensor for inhomogeneous fluids. *Phys. Rev. E*, 52 (2):1627–1638, 1995. Cited on pages 6, 77, and 78.

[5] F. Varnik, J. Baschnagel, and K. Binder. Molecular dynamics results on the pressure tensor of polymer films. *J. Chem. Phys.*, 113:4444, 2000. Cited on pages 6, 74, and 77.

[6] S. Chapman and T.G. Cowling. *The Mathematical Theory of Non-uniform Gases*. Cambridge University Press, 1952. Cited on pages 6 and 133.

[7] G. A. Truskey, F. Yuan, and D. F. Katz. *Transport phenomena in biological systems*. Pearson/Prentice Hall, 2004. Cited on pages 6, 12, 13, and 14.

[8] T.E. Faber. *Fluid dynamics for physicists*. Cambridge University Press, 1995. ISBN 0521429692. Cited on pages 6, 8, and 106.

[9] R. Pal. Rheology of simple and multiple emulsions. *Curr. Opin. Colloid In.*, 16:41–60, 2011. Cited on pages 7, 9, and 10.

[10] G.K. Batchelor. The stress system in a suspension of force-free particles. *J. Fluid Mech.*, 41(03): 545–570, 1970. Cited on pages 7, 8, and 75.

[11] J.M. Brader. Nonlinear rheology of colloidal dispersions. *J. Phys. Condens. Matter*, 22:1–36, 2010. Cited on pages 7, 8, 9, and 93.

[12] C.K. Aidun and J.R. Clausen. Lattice-Boltzmann Method for Complex Flows. *Annu. Rev. Fluid Mech.*, 42:439–472, 2010. Cited on pages 7, 22, and 29.

[13] J.K.G. Dhont. *An Introduction to Dynamics of Colloids*. Elsevier, 1996. Cited on pages 7, 8, 19, and 107.

[14] A. Einstein. A new determination of molecular dimensions. *Ann. Phys.*, 19:289–306, 1906. Cited on pages 8 and 74.

[15] G. Bossis and J.F. Brady. The rheology of Brownian suspensions. *J. Chem. Phys.*, 91:1866, 1989. Cited on page 8.

[16] H.A. Barnes. Shear-thickening (dilatancy) in suspensions of nonaggregating solid particles dispersed in Newtonian liquids. *J. Rheol.*, 33:329, 1989. Cited on page 8.

[17] T. Dabak and O. Yucel. Modeling of the concentration particle size distribution effects on the rheology of highly concentrated suspensions. *Powder Technol.*, 52(3):193–206, 1987. Cited on page 9.

[18] L. Heymann, S. Peukert, and N. Aksel. On the solid-liquid transition of concentrated suspensions in transient shear flow. *Rheol. Acta*, 41(4):307–315, 2002. Cited on page 9.

[19] M. Fuchs and M.E. Cates. Schematic models for dynamic yielding of sheared colloidal glasses. *Faraday Discuss.*, 123:267–286, 2002. Cited on pages 9 and 93.

[20] F. Varnik, L. Bocquet, and J.L. Barrat. A study of the static yield stress in a binary Lennard-Jones glass. *J. Chem. Phys.*, 120:2788, 2004. Cited on pages 9 and 31.

[21] F. Varnik and D. Raabe. Profile blunting and flow blockage in a yield-stress fluid: a molecular dynamics study. *Phys. Rev. E*, 77:011504, 2008. Cited on pages 9 and 31.

[22] J.-L. Barrat and A. Lemaître. *Heterogeneities in amorphous systems under shear*, chapter 8. Oxford University Press, 2011. Cited on page 9.

[23] T.W. Secomb, S. Chien, K.M. Jan, and R. Skalak. The bulk rheology of close-packed red blood cells in shear flow. *Biorheology*, 20(3):295, 1983. Cited on page 9.

[24] T.W. Secomb, T.M. Fischer, and R. Skalak. The motion of close-packed red blood cells in shear flow. *Biorheology*, 20(3):283, 1983. Cited on page 9.

[25] P. Bagchi and R.M. Kalluri. Rheology of a dilute suspension of liquid-filled elastic capsules. *Phys. Rev. E*, 81(5):056320, 2010. Cited on pages 9 and 76.

[26] D. Barthès-Biesel and V. Chhim. The constitutive equation of a dilute suspension of spherical microcapsules. *Int. J. Multiphase Flow*, 7(5):493–505, 1981. Cited on page 9.

[27] D. Barthès-Biesel and A. Acrivos. The rheology of suspensions and its relation to phenomenological theories for non-Newtonian fluids. *Int. J. Multiphase Flow*, 1(1):1–24, 1973. Cited on page 9.

[28] M. Bredimas, M. Veyssie, D. Barthès-Biesel, and V. Chhim. Model suspension of spherical capsules: Physical and rheological properties. *J. Colloid Interface Sci.*, 93(2):513–520, 1983. Cited on page 10.

[29] M. Bessis and R.I. Weed. *Living blood cells and their ultrastructure*. Springer, 1973. Cited on page 11.

[30] T.M. Fischer, M. Stöhr-Liesen, and H. Schmid-Schönbein. The Red Cell as a Fluid Droplet: Tank Tread-Like Motion of the Human Erythrocyte Membrane in Shear Flow. *Science*, 202:894–896, 1978. Cited on pages 11, 12, and 13.

[31] G. Gompper and M. Schick. *Soft Matter: Lipid Bilayers and Red Blood Cells*. Wiley-VCH, 2008. Cited on pages 11, 12, 18, 19, 43, 45, 46, 47, 48, 65, and 85.

[32] R. Skalak and P.I. Branemark. Deformation of red blood cells in capillaries. *Science*, 164:717–719, 1969. Cited on page 11.

[33] N. Mohandas and E. Evans. Mechanical properties of the red cell membrane in relation to molecular structure and genetic defects. *Annu. Rev. Bioph. Biom.*, 23:787–818, 1994. Cited on page 11.

[34] M. Dao, C.T. Lim, and S. Suresh. Mechanics of the human red blood cell deformed by optical tweezers. *J. Mech. Phys. Solids*, 51(11-12):2259–2280, 2003. Cited on page 11.

[35] S. Svetina, D. Kuzman, R.E. Waugh, P. Ziherl, and B. Zeks. The Cooperative Role of Membrane Skeleton and Bilayer in the Mechanical Behaviour of Red Blood Cells. *Bioelectroch.*, 62(2):107–113, 2004. Cited on pages 11 and 46.

[36] E.A. Evans. Bending Resistance and Chemically Induced Moments in Membrane Bilayers. *Biophys. J.*, 14(12):923–931, 1974. Cited on pages 11 and 46.

[37] E.A. Evans and R.M. Hochmuth. Membrane viscoelasticity. *Biophys. J.*, 16(1):1–11, 1976. Cited on page 11.

[38] E.A. Evans and R. Skalak. *Mechanics and thermodynamics of biomembranes*. CRC, 1980. Cited on pages 11 and 47.

[39] M. de Oliveira, C. Vera, P. Valdez, Y. Sharma, R. Skelton, and L.A. Sung. Nanomechanics of Multiple Units in the Erythrocyte Membrane Skeletal Network, 2010. URL `http://www.springerlink.com/content/82741n233x574637/`. Cited on page 12.

[40] G. Gompper and D. Kroll. *Triangulated-surface models of fluctuating membranes*, pages 1–71. World Scientific, 2004. Cited on pages 12 and 46.

[41] E.A. Evans. Bending elastic modulus of red blood cell membrane derived from buckling instability in micropipet aspiration tests. *Biophys. J.*, 43(1):27–30, 1983. Cited on page 12.

[42] L. Scheffer, A. Bitler, E. Ben-Jacob, and R. Korenstein. Atomic force pulling: probing the local elasticity of the cell membrane. *Eur. Biophys. J.*, 30(2):83–90, 2001. Cited on page 12.

[43] C.Y. Chee, H.P. Lee, and C. Lu. Using 3D fluid-structure interaction model to analyse the biomechanical properties of erythrocyte. *Phys. Lett. A*, 372(9):1357–1362, 2008. Cited on pages 12 and 126.

[44] A. Ghosh, S. Sinha, J.A. Dharmadhikari, S. Roy, A.K. Dharmadhikari, J. Samuel, S. Sharma, and D. Mathur. Euler buckling-induced folding and rotation of red blood cells in an optical trap. *Phys. Biol.*, 3(1):67–73, 2006. Cited on pages 12 and 46.

[45] G. Lenormand, S. Hénon, A. Richert, J. Siméon, and F. Gallet. Elasticity of the human red blood cell skeleton. *Biorheology*, 40(1):247–251, 2003. Cited on page 12.

[46] A. Robertson, A. Sequeira, and M. Kameneva. *Hemorheology*, pages 63–120. Oberwolfach Seminars. Birkhäuser Basel, 2007. Cited on pages 12, 13, 14, 32, 97, and 123.

[47] H. Schmid-Schönbein and R. Wells. Fluid Drop-Like Transition of Erythrocytes under Shear. *Science*, 165(3890):288–291, 1969. Cited on pages 12, 13, 14, and 18.

[48] C. Pozrikidis. Axisymmetric motion of a file of red blood cells through capillaries. *Phys. Fluids*, 17: 031503, 2005. Cited on pages 12 and 13.

[49] S. Suresh, J. Spatz, J.P. Mills, A. Micoulet, M. Dao, C.T. Lim, M. Beil, and M. Seufferlein. Connections between single-cell biomechanics and human disease states: gastrointestinal cancer and malaria. *Acta Biomaterialia*, 1:15–30, 2005. Cited on page 12.

[50] G.A. Barabino, M.O. Platt, and D.K. Kaul. Sickle Cell Biomechanics. *Annu. Rev. Biomed. Eng.*, 12:345–367, 2010. Cited on pages 12 and 13.

[51] A.S. Popel and P.C. Johnson. Microcirculation and Hemorheology. *Annu. Rev. Fluid Mech.*, 37: 43–69, 2005. Cited on pages 12, 13, 14, 19, and 32.

[52] T. AlMomani, H.S. Udaykumar, J.S. Marshall, and K.B. Chandran. Micro-scale Dynamic Simulation of Erythrocyte–Platelet Interaction in Blood Flow. *Ann. Biomed. Eng.*, 36(6):905–920, 2008. Cited on pages 13 and 125.

[53] D.M. Wootton and D.N. Ku. Fluid Mechanics of Vascular Systems, Diseases, and Thrombosis. *Annu. Rev. Biomed. Eng.*, 1:299–329, 1999. Cited on pages 13 and 32.

[54] S. Chien. Shear Dependence of Effective Cell Volume as a Determinant of Blood Viscosity. *Science*, 168(3934):977–979, 1970. Cited on pages 13, 18, 97, and 122.

[55] H. Schmid-Schönbein. Microrheology of erythrocytes, blood viscosity, and the distribution of blood flow in the microcirculation. *Internat. Rev. Physiol. Cardiovasc. Physiol. II*, 9:1–59, 1976. Cited on page 13.

[56] S. Chien and K. Jan. Ultrastructural Basis of the Mechanism of Rouleaux Formation. *Microvasc. Res.*, 5(2):155–166, 1973. Cited on page 13.

[57] R. Skalak, P.R. Zarda, K.-M. Jan, and S. Chien. Mechanics of rouleau formation. *Biophys. J.*, 35 (3):771–781, 1981. Cited on page 13.

[58] L. Game, J.C. Voegel, P. Schaaf, and J.F. Stoltz. Do physiological concentrations of IgG induce a direct aggregation of red blood cells: comparison with fibrinogen. *Biochimica Biophysica Acta*, 1291 (2):138–142, 1996. ISSN 0304-4165. Cited on page 13.

[59] J.J. Bishop, A.S. Popel, M. Intaglietta, and P.C. Johnson. Rheological effects of red blood cell aggregation in the venous network: a review of recent studies. *Biorheology*, 38(2):263–274, 2001. ISSN 0006-355X. Cited on page 13.

[60] S. Chien, R.G. King, R. Skalak, S. Usami, and A.L. Copley. Viscoelastic properties of human blood and red cell suspensions. *Biorheology*, 12(6):341–346, 1975. ISSN 0006-355X. Cited on page 13.

[61] E.W. Merrill, E.R. Gilliland, G. Cokelet, H. Shin, A. Britten, and R.E. Wells Jr. Rheology of Human Blood, near and at Zero Flow. Effects of Temperature and Hematocrit Level. *Biophys. J.*, 3(3): 199–213, 1963. ISSN 0006-3495. Cited on page 13.

[62] S.E. Charm and G.S. Kurland. Static Method for determining Blood Yield Stress. *Nature*, 216: 1121–1123, 1967. Cited on page 14.

[63] C. Picart, J.-M. Piau, H. Galliard, and P. Carpentier. Human blood shear yield stress and its hematocrit dependence. *J. Rheol.*, 42:1–12, 1998. Cited on page 14.

[64] J. Zhang, P.C. Johnson, and A.S. Popel. Red Blood Cell Aggregation and Dissociation in Shear Flows Simulated by Lattice Boltzmann Method. *J. Biomech.*, 41(1):47–55, 2008. Cited on pages 14 and 22.

[65] T. Secomb, B. Styp-Rekowska, and A. Pries. Two-Dimensional Simulation of Red Blood Cell Deformation and Lateral Migration in Microvessels. *Ann. Biomed. Eng.*, 35(5):755–765, 2007. Cited on page 14.

[66] H.L. Goldsmith. Red cell motions and wall interactions in tube flow. *Fed. Proc.*, 30:1578–1588, 1971. Cited on page 14.

[67] H. Brenner and J. Happel. *Low Reynolds number hydrodynamics*. Kluwer, 1983. Cited on page 14.

[68] J. Zhang, P.C. Johnson, and A.S. Popel. Effects of erythrocyte deformability and aggregation on the cell free layer and apparent viscosity of microscopic blood flows. *Microvasc. Res.*, 77:265–272, 2009. Cited on pages 14 and 17.

[69] H.L. Goldsmith, G.R. Cokelet, and P. Gaehtgens. Robin Fåhraeus: evolution of his concepts in cardiovascular physiology. *Am. J. Physiol.*, 257(3):H1005, 1989. ISSN 0363-6135. Cited on page 14.

[70] R. Fåhraeus. The suspension stability of the blood. *Physiol. Rev.*, 9(2):241–274, 1929. ISSN 0031-9333. Cited on page 14.

[71] R. Fåhraeus and T. Lindqvist. The viscosity of blood in narrow capillary tubes. *Am. J. Physiol.*, 96:562–568, 1931. Cited on page 14.

[72] G.J. Tangelder, D.W. Slaaf, A.M.M. Muijtjens, T. Arts, M.G.A. oude Egbrink, R.S. Reneman, et al. Velocity Profiles of Blood Platelets and Red Blood Cells Flowing in Arterioles of the Rabbit Mesentery. *Circ. Res.*, 59(5):505–514, 1986. Cited on page 14.

[73] A.J.C. Ladd. Numerical simulations of particulate suspensions via a discretized Boltzmann equation. Part 1. Theoretical foundation. *J. Fluid Mech.*, 271:285–309, 1994. Cited on pages 17, 22, 24, 29, 30, 32, and 66.

[74] A.J.C. Ladd. Numerical simulations of particulate suspensions via a discretized Boltzmann equation. Part 2. Numerical results. *J. Fluid Mech.*, 271:311–339, 1994. Cited on page 17.

[75] C.K. Aidun and Y. Lu. Lattice Boltzmann simulation of solid particles suspended in fluid. *J. Stat. Phys.*, 81(1):49–61, 1995. Cited on pages 17 and 31.

[76] M. Kraus, W. Wintz, U. Seifert, and R. Lipowsky. Fluid Vesicles in Shear Flow. *Phys. Rev. Lett.*,

77(17):3685–3688, 1996. Cited on pages 17, 18, 46, 52, and 90.

[77] C.D. Eggleton and A.S. Popel. Large Deformation of Red Blood Cell Ghosts in a Simple Shear Flow. *Phys. Fluids*, 10(8):1834–1845, 1998. Cited on pages 17, 38, 41, 47, and 65.

[78] C.S. Peskin. The Immersed Boundary Method. *Acta Numerica*, 11:479–517, 2002. Cited on pages 17, 37, 38, 39, 40, 41, and 64.

[79] Z.-G. Feng and E.E. Michaelides. The Immersed Boundary-Lattice Boltzmann Method for Solving Fluid-Particles Interaction Problems. *J. Comput. Phys.*, 195(2):602–628, 2004. Cited on pages 17, 22, 63, 65, 66, 67, and 69.

[80] J. Zhang, P.C. Johnson, and A.S. Popel. An Immersed Boundary Lattice Boltzmann Approach to Simulate Deformable Liquid Capsules and its Application to Microscopic Blood Flows. *Phys. Biol.*, 4(4):285–295, 2007. Cited on pages 17, 22, and 38.

[81] C. Sun and L.L. Munn. Particulate nature of blood determines macroscopic rheology: a 2-D lattice Boltzmann analysis. *Biophys. J.*, 88(3):1635–1645, 2005. Cited on page 17.

[82] C. Sun and L.L. Munn. Influence of erythrocyte aggregation on leukocyte margination in postcapillary expansions: a lattice Boltzmann analysis. *Physica A*, 362(1):191–196, 2006. Cited on page 17.

[83] H. Noguchi and G. Gompper. Fluid Vesicles with Viscous Membranes in Shear Flow. *Phys. Rev. Lett.*, 93(25):258102, 2004. Cited on pages 17 and 126.

[84] C. Pozrikidis. Effect of Membrane Bending Stiffness on the Deformation of Capsules in Simple Shear Flow. *J. Fluid Mech.*, 440:269–291, 2001. Cited on pages 17, 46, 47, 65, and 101.

[85] C. Pozrikidis. Numerical simulation of cell motion in tube flow. *Ann. Biomed. Eng.*, 33(2):165–178, 2005. Cited on pages 17 and 63.

[86] C. Pozrikidis. Numerical Simulation of the Flow-Induced Deformation of Red Blood Cells. *Ann. Biomed. Eng.*, 31(10):1194–1205, 2003. Cited on pages 17, 46, and 101.

[87] M.M. Dupin, I. Halliday, C.M. Care, L. Alboul, and L.L. Munn. Modeling the Flow of Dense Suspensions of Deformable Particles in Three Dimensions. *Phys. Rev. E*, 75(6):066707, 2007. Cited on pages 17, 47, 61, and 90.

[88] P. Bagchi. Mesoscale Simulation of Blood Flow in Small Vessels. *Biophys. J.*, 92(6):1858–1877, 2007. Cited on pages 17, 63, and 126.

[89] S.K. Doddi and P. Bagchi. Three-dimensional computational modeling of multiple deformable cells flowing in microvessels. *Phys. Rev. E*, 79(4):046318, 2009. Cited on pages 17, 38, 46, 63, and 126.

[90] R.M. MacMeccan, J.R. Clausen, G.P. Neitzel, and C.K. Aidun. Simulating deformable particle suspensions using a coupled lattice-Boltzmann and finite-element method. *J. Fluid Mech.*, 618: 13–39, 2009. Cited on page 17.

[91] S. Shin, Y. Ku, M.S. Park, and J.S. Suh. Measurement of red cell deformability and whole blood viscosity using laser-diffraction slit rheometer. *Korea-Australia Rheology Journal*, 16(2):85–90, 2004. Cited on page 18.

[92] T.M. Fischer. Shape Memory of Human Red Blood Cells. *Biophys. J.*, 86:3304–3313, 2004. Cited on pages 19 and 47.

[93] C. Pozrikidis, editor. *Modeling and Simulation of Capsules and Biological Cells.* Chapman & Hall/CRC Mathematical Biology and Medicine Series, 2003. Cited on pages 19 and 126.

[94] V. Breedveld. *Shear-induced self-diffusion in concentrated suspensions.* PhD thesis, University of Twente, 2000. Cited on pages 19, 107, 110, 111, 112, 113, and 124.

[95] T. Powers. Dynamics of filaments and membranes in a viscous fluid. *Rev. Modern Phys.*, 82: 1607–1631, 2010. Cited on page 19.

[96] U. Frisch, B. Hasslacher, and Y. Pomeau. Lattice Gas Automata for the Navier-Stokes Equation. *Phys. Rev. Lett.*, 56(14):1505–1508, 1986. Cited on page 21.

[97] S. Wolfram. Cellular automaton fluids 1: Basic theory. *J. Stat. Phys.*, 45(3):471–526, 1986. ISSN 0022-4715. Cited on page 21.

[98] S. Succi. *The Lattice Boltzmann Equation for Fluid Dynamics and Beyond.* Oxford University Press, 2001. ISBN 978-0198503989. Cited on pages 21, 22, and 36.

[99] D. Hänel. *Molekulare Gasdynamik.* Springer, 2004. Cited on pages 21 and 133.

[100] G.R. McNamara and G. Zanetti. Use of the Boltzmann Equation to Simulate Lattice-Gas Automata. *Phys. Rev. Lett.*, 61(20):2332–2335, 1988. Cited on page 21.

[101] F.J. Higuera and J. Jiménez. Boltzmann Approach to Lattice Gas Simulations. *Europhys. Lett.*, 9 (7):663–668, 1989. Cited on page 21.

[102] F.J. Higuera, S. Succi, and R. Benzi. Lattice gas dynamics with enhanced collisions. *Europhys. Lett.*, 9:345, 1989. Cited on page 21.

[103] P.L. Bhatnagar, E.P. Gross, and M. Krook. A model for collision processes in gases. I. Small

amplitude processes in charged and neutral one-component systems. *Phys. Rev.*, 94(3):511–525, 1954. Cited on page 21.

[104] Y.H. Qian. *Lattice Gas and lattice kinetic theory applied to Navier-Stokes equation.* PhD thesis, Université Pierre et Marie Curie, Paris, 1990. Cited on page 21.

[105] H. Chen, D. Chen, and W. Matthaeus. Recovery of the Navier-Stokes Equations through a Lattice Gas Boltzmann Equation Method. *Phys. Rev. A*, 45:5339–5342, 1992. Cited on page 21.

[106] Y.-H. Qian, D. d'Humières, and P. Lallemand. Lattice BGK Models for Navier-Stokes Equation. *Europhys. Lett.*, 17:479–484, 1992. Cited on pages 21, 24, and 25.

[107] X. He and L.-S. Luo. A priori derivation of the lattice Boltzmann equation. *Phys. Rev. E*, 55(6): 6333–6336, 1997. Cited on page 22.

[108] T. Abe. Derivation of the lattice Boltzmann method by means of the discrete ordinate method for the Boltzmann equation. *J. Comput. Phys.*, 131(1):241–246, 1997. Cited on page 22.

[109] S. Chen and G.D. Doolen. Lattice Boltzmann Method for Fluid Flows. *Annu. Rev. Fluid Mech.*, 30: 329–364, 1998. Cited on page 22.

[110] A.J.C. Ladd and R. Verberg. Lattice-Boltzmann simulations of particle-fluid suspensions. *J. Stat. Phys.*, 104(5):1191–1251, 2001. Cited on pages 22, 24, 25, 29, 30, 66, 133, and 134.

[111] E.S. Boek and M. Venturoli. Lattice-Boltzmann studies of fluid flow in porous media with realistic rock geometries. *Comput. Math. Appl.*, 59(7):2305–2314, 2010. Cited on page 22.

[112] Q. Kang, P.C. Lichtner, and D.R. Janecky. Lattice Boltzmann Method for Reacting Flows in Porous Media. *Adv. Appl. Math. Mech.*, 2(5):545–563, 2010. Cited on page 22.

[113] D. Sun, M. Zhu, S. Pan, and D. Raabe. Lattice Boltzmann modeling of dendritic growth in a forced melt convection. *Acta Mater.*, 57(6):1755–1767, 2009. Cited on page 22.

[114] R. Benzi and S. Succi. Two-dimensional turbulence with the lattice Boltzmann equation. *J. Phys. A*, 23:L1, 1990. Cited on page 22.

[115] S. Hou, J. Sterling, S. Chen, and G.D. Doolen. A lattice Boltzmann subgrid model for high Reynolds number flows. *Fields Inst. Commun.*, 6:151–166, 1996. Cited on page 22.

[116] M. Mendoza and J.D. Muñoz. Three-dimensional lattice Boltzmann model for electrodynamics. *Phys. Rev. E*, 82:056708, 2010. Cited on page 22.

[117] M. Mendoza, B.M. Boghosian, H.J. Herrmann, and S. Succi. Fast Lattice Boltzmann Solver for Relativistic Hydrodynamics. *Phys. Rev. Lett.*, 105(1):014502, 2010. Cited on page 22.

[118] D.A. Wolf-Gladrow. *Lattice-Gas Cellular Automata and Lattice Boltzmann Models - An Introduction.* Lecture Notes in Mathematics. Springer, 2000. Cited on pages 22 and 133.

[119] M. Sukop and D. Thorne. *Lattice Boltzmann Modeling, an Introduction for Geoscientists and Engineers.* Springer, 2005. ISBN 978-3540279815. Cited on page 22.

[120] L.-S. Luo. The lattice-gas and lattice Boltzmann methods: past, present, and future. In *Proceedings of the International Conference on Applied Computational Fluid Dynamics, Beijing, China*, pages 52–83, 2000. Cited on page 22.

[121] D. Yu, R. Mei, L.-S. Luo, and W. Shyy. Viscous flow computations with the method of lattice Boltzmann equation. *Prog. Aerosp. Sci.*, 39(5):329–367, 2003. Cited on page 22.

[122] D. Raabe. Overview of the Lattice Boltzmann Method for Nano- and Microscale Fluid Dynamics in Materials Science and Engineering. *Model. Simul. Mater. Sc.*, 12:R13–R46, 2004. Cited on page 22.

[123] B. Chopard and M. Droz. *Cellular automata modeling of physical systems.* Cambridge University Press, 1998. ISBN 0521461685. Cited on page 24.

[124] Z. Guo, C. Zheng, and B. Shi. Discrete Lattice Effects on the Forcing Term in the Lattice Boltzmann Method. *Phys. Rev. E*, 65(4):046308, 2002. Cited on pages 24, 25, 133, and 134.

[125] M. Gross, N. Moradi, G. Zikos, and F. Varnik. Shear stress in nonideal fluid lattice Boltzmann simulations. *Phys. Rev. E*, 83:017701, 2011. Cited on page 25.

[126] J. Latt. *Hydrodynamic Limit of Lattice Boltzmann Equations.* PhD thesis, University of Geneva, 2007. Cited on pages 26, 27, 28, 36, 137, and 144.

[127] P.A. Skordos. Initial and Boundary Conditions for the Lattice Boltzmann Method. *Phys. Rev. E*, 48(6):4823–4842, 1993. Cited on pages 27 and 28.

[128] R. Mei, L.-S. Luo, P. Lallemand, and D. d'Humières. Consistent initial conditions for lattice Boltzmann simulations. *Comput. Fluids*, 35(8-9):855–862, 2006. Cited on pages 27 and 141.

[129] C. Pozrikidis. *Fluid Dynamics: Theory, Computation, and Numerical Simulation.* Springer, 2nd edition, 2009. ISBN 978-0-387-95869-9. Cited on pages 27 and 76.

[130] J.C. Van der Werff, C.G. De Kruif, C. Blom, and J. Mellema. Linear viscoelastic behavior of dense hard-sphere dispersions. *Phys. Rev. A*, 39(2):795–807, 1989. Cited on page 27.

[131] M.A. Gallivan, D.R. Noble, J.G. Georgiadis, and R.O. Buckius. An evaluation of the bounce-back

boundary condition for lattice Boltzmann simulations. *Int. J. Numer. Meth. Fluids*, 25(3):249–263, 1997. Cited on pages 28 and 29.

[132] J. Latt, B. Chopard, O. Malaspinas, M. Deville, and A. Michler. Straight Velocity Boundaries in the Lattice Boltzmann Method. *Phys. Rev. E*, 77(5):056703, 2008. Cited on page 28.

[133] Q. Zou and X. He. On pressure and velocity boundary conditions for the lattice Boltzmann BGK model. *Phys. Fluids*, 9(6):1591–1598, 1997. Cited on page 28.

[134] J.C.G. Verschaeve. Analysis of the lattice Boltzmann Bhatnagar-Gross-Krook no-slip boundary condition: Ways to improve accuracy and stability. *Phys. Rev. E*, 80(3):36703, 2009. Cited on page 28.

[135] T. Inamuro, M. Yoshino, and F. Ogino. A non-slip boundary condition for lattice Boltzmann simulations. *Phys. Fluids*, 7(12):2928–2930, 1995. Cited on page 28.

[136] J.C.G. Verschaeve and B. Müller. A curved no-slip boundary condition for the lattice Boltzmann method. *J. Comput. Phys.*, 229:6781–6803, 2010. Cited on page 28.

[137] Z. Yang. Pressure Conditions for lattice Boltzmann methods on domains with curved boundaries. *Comput. Math. Appl.*, 59:2168–2177, 2010. Cited on page 28.

[138] K. Mattila, J. Hyväluoma, A.A. Folarin, and T. Rossi. A boundary condition for arbitrary shaped inlets in lattice-Boltzmann simulations. *Int. J. Numer. Meth. Fluids*, 63(5):638–650, 2010. Cited on page 28.

[139] M. Junk and Z. Yang. Outflow boundary conditions for the lattice Boltzmann method. *Prog. Comput. Fluid Dyn.*, 8(1):38–48, 2008. Cited on page 28.

[140] S. Izquierdo, P. Martínez-Lera, and N. Fueyo. Analysis of open boundary effects in unsteady lattice Boltzmann simulations. *Comput. Math. Appl.*, 58(5):914–921, 2009. Cited on page 28.

[141] L. Szalmás. Slip-flow boundary conditions for straight walls in the lattice Boltzmann model. *Phys. Rev. E*, 73:066710, 2006. Cited on page 28.

[142] S. Izquierdo and N. Fueyo. Momentum transfer correction for macroscopic-gradient boundary conditions in lattice Boltzmann methods. *J. Comput. Phys.*, 229:2497–2506, 2010. Cited on pages 28 and 31.

[143] S.H. Kim and H. Pitsch. A generalized periodic boundary condition for lattice Boltzmann method simulation of a pressure driven flow in a periodic geometry. *Phys. Fluids*, 19(10):108101, 2007. Cited on page 28.

[144] A.J. Wagner and I. Pagonabarraga. Lees–Edwards Boundary Conditions for Lattice Boltzmann. *J. Stat. Phys.*, 107(1):521–537, 2002. Cited on pages 28, 31, and 126.

[145] E. Lorenz, A.G. Hoekstra, and A. Caiazzo. Lees-Edwards boundary conditions for lattice Boltzmann suspension simulations. *Phys. Rev. E*, 79(3):036706, 2009. Cited on pages 28, 31, and 126.

[146] R. Cornubert, D. d'Humières, and D. Levermore. A Knudsen layer theory for lattice gases. *Physica D*, 47(1-2):241–259, 1991. Cited on page 29.

[147] I. Ginzbourg and P.M. Adler. Boundary flow condition analysis for the three-dimensional lattice Boltzmann model. *J. Phys. II France*, 4(2):191–214, 1994. Cited on page 29.

[148] M. Bouzidi, M. Firdaouss, and P. Lallemand. Momentum transfer of a Boltzmann-lattice fluid with boundaries. *Phys. Fluids*, 13(11):3452–3459, 2001. Cited on pages 29 and 31.

[149] D. d'Humières, M. Bouzidi, and P. Lallemand. Thirteen-velocity three-dimensional lattice Boltzmann model. *Phys. Rev. E*, 63(6):066702, 2001. Cited on pages 29 and 141.

[150] X. He, Q. Zou, L.-S. Luo, and M. Dembo. Analytic Solutions of Simple Flows and Analysis of Nonslip Boundary Conditions for the Lattice Boltzmann BGK Model. *J. Stat. Phys.*, 87(1):115–136, 1997. Cited on pages 29, 31, and 35.

[151] O. Filippova and D. Hänel. Grid refinement for lattice-BGK models. *J. Comput. Phys.*, 147(1): 219–228, 1998. Cited on page 29.

[152] R. Mei, D. Yu, W. Shyy, and L.-S. Luo. Force Evaluation in the Lattice Boltzmann Method Involving Curved Geometry. *Phys. Rev. E*, 65(4):041203, 2002. Cited on page 29.

[153] D.P. Ziegler. Boundary Conditions for Lattice Boltzmann Simulations. *J. Stat. Phys.*, 71(5): 1171–1177, 1993. Cited on page 30.

[154] E.-J. Ding and C.K. Aidun. Extension of the Lattice-Boltzmann Method for Direct Simulation of Suspended Particles Near Contact. *J. Stat. Phys.*, 112(3):685–708, 2003. Cited on pages 31, 51, and 66.

[155] A.M. Artoli, D. Kandhai, H.C.J. Hoefsloot, A.G. Hoekstra, and P.M.A. Sloot. Lattice BGK Simulations of Flow in a Symmetric Bifurcation. *Future Gener. Comp. Sy.*, 20(6):909–916, 2004. Cited on page 31.

[156] F. Janoschek, F. Toschi, and J. Harting. A simplified particulate model for coarse-grained hemody-

namics simulations. *Phys. Rev. E*, 82:056710, 2010. Cited on page 31.
[157] O. Gräser and A. Grimm. Adaptive generalized periodic boundary conditions for lattice Boltzmann simulations of pressure-driven flows through confined repetitive geometries. *Phys. Rev. E*, 82:016702, 2010. Cited on page 31.
[158] F. Varnik, D. Dorner, and D. Raabe. Roughness-Induced Flow Instability: A Lattice Boltzmann Study. *J. Fluid Mech.*, 573:191–209, 2007. Cited on page 31.
[159] Y. Miyazaki, S. Nomura, T. Miyake, H. Kagawa, C. Kitada, H. Taniguchi, Y. Komiyama, Y. Fujimura, Y. Ikeda, and S. Fukuhara. High shear stress can initiate both platelet aggregation and shedding of procoagulant containing microparticles. *Blood*, 88(9):3456, 1996. Cited on page 32.
[160] S.A. Berger and L.-D. Jou. Flows in Stenotic Vessels. *Annu. Rev. Fluid Mech.*, 32(1):347–382, 2000. Cited on page 32.
[161] J.J. Hathcock. Flow Effects on Coagulation and Thrombosis. *Arterioscl. Throm. Vas. Biol.*, 26(8): 1729–1737, 2006. Cited on page 32.
[162] S.W. Schneider, S. Nuschele, A. Wixforth, C. Gorzelanny, A. Alexander-Katz, R.R. Netz, and M.F. Schneider. Shear-induced unfolding triggers adhesion of von Willebrand factor fibers. *P. Natl. Acad. Sci. USA*, 104(19):7899–7903, 2007. Cited on page 32.
[163] J.D. Sterling and S. Chen. Stability Analysis of Lattice Boltzmann Methods. *J. Comput. Phys.*, 123:196–206, 1996. Cited on page 35.
[164] D.J. Holdych, D.R. Noble, J.G. Georgiadis, and R.O. Buckius. Truncation Error Analysis of Lattice Boltzmann Methods. *J. Comput. Phys.*, 193(2):595–619, 2004. Cited on pages 35, 36, 137, 138, 139, and 141.
[165] T. Krüger, F. Varnik, and D. Raabe. Shear stress in lattice Boltzmann simulations. *Phys. Rev. E*, 79(4):046704, 2009. Cited on pages 35, 36, 122, 138, 139, 140, and 141.
[166] Y.T. Feng, K. Han, and D.R.J. Owen. Coupled lattice Boltzmann method and discrete element modelling of particle transport in turbulent fluid flows: Computational issues. *Int. J. Numer. Meth. Eng.*, 72(9):1111–1134, 2007. Cited on pages 36 and 51.
[167] C.S. Peskin. *Flow patterns around heart valves: A digital computer method for solving the equations of motion.* Sue Golding Graduate Division of Medical Sciences, Albert Einstein College of Medicine, Yeshiva University, 1972. Cited on pages 37 and 38.
[168] C.S. Peskin. Numerical analysis of blood flow in the heart. *J. Comput. Phys.*, 25(3):220–252, 1977. Cited on page 37.
[169] Y. Sui, Y.T. Chew, P. Roy, and H.T. Low. A hybrid method to study flow-induced deformation of three-dimensional capsules. *J. Comput. Phys.*, 227(12):6351–6371, 2008. Cited on pages 38 and 45.
[170] A.L. Fogelson and R.D. Guy. Immersed-boundary-type models of intravascular platelet aggregation. *Comput. Methods Appl. Mech. Engrg.*, 197(25-28):2087–2104, 2008. Cited on page 38.
[171] C. Tu and C.S. Peskin. Stability and Instability in the Computation of Flows with Moving Immersed Boundaries: A Comparison of Three Methods. *SIAM J. Sci. Stat. Comp.*, 13(6):1361–1376, 1992. Cited on page 38.
[172] C.S. Peskin and B.F. Printz. Improved Volume Conservation in the Computation of Flows with Immersed Elastic Boundaries. *J. Comput. Phys.*, 105:33–46, 1993. Cited on pages 38 and 48.
[173] R.J. LeVeque and Z. Li. Immersed interface methods for Stokes flow with elastic boundaries or surface tension. *SIAM J. Sci. Stat. Comp.*, 18(3):709–735, 1997. Cited on page 38.
[174] A.M. Roma, C.S. Peskin, and M.J. Berger. An adaptive version of the immersed boundary method. *J. Comput. Phys.*, 153(2):509–534, 1999. Cited on page 38.
[175] D. McQueen and C.S. Peskin. Shared-memory parallel vector implementation of the immersed boundary method for the computation of blood flow in the beating mammalian heart. *J. Supercomput.*, 11(3):213–236, 1997. Cited on page 38.
[176] A.L. Fogelson and C.S. Peskin. A fast numerical method for solving the three-dimensional Stokes' equations in the presence of suspended particles. *J. Comput. Phys.*, 79(1):50–69, 1988. Cited on page 38.
[177] N.T. Wang and A.L. Fogelson. Computational Methods for Continuum Models of Platelet Aggregation. *J. Comput. Phys.*, 151(2):649–675, 1999. Cited on page 38.
[178] C.C. Vesier and A.P. Yoganathan. A computer method for simulation of cardiovascular flow fields: Validation of approach. *J. Comput. Phys.*, 99(2):271–287, 1992. Cited on page 38.
[179] R. Dillon, L. Fauci, A.L. Fogelson, and D. Gaver III. Modeling biofilm processes using the immersed boundary method. *J. Comput. Phys.*, 129(1):57–73, 1996. Cited on page 38.
[180] M.-C. Lai and C.S. Peskin. An Immersed Boundary Method with Formal Second-Order Accuracy and Reduced Numerical Viscosity. *J. Comput. Phys.*, 160:705–719, 2000. Cited on page 38.

[181] R. Mittal and G. Iaccarino. Immersed boundary methods. *Annu. Rev. Fluid Mech.*, 37:239–261, 2005. Cited on page 38.

[182] B. Dünweg and A.J.C. Ladd. *Lattice Boltzmann Simulations of Soft Matter Systems*, volume 221 of *Advanced Computer Simulation Approaches for Soft Matter Sciences III, Advances in Polymer Science*, page 89. Springer, 2009. Cited on pages 41, 42, 64, and 133.

[183] G. Tryggvason, B. Bunner, A. Esmaeeli, D. Juric, N. Al-Rawahi, W. Tauber, J. Han, S. Nas, and Y.-J. Jan. A front-tracking method for the computations of multiphase flow. *J. Comput. Phys.*, 169 (2):708–759, 2001. Cited on pages 41 and 126.

[184] N. Peller, A. Le Duc, F. Tremblay, and M. Manhart. High-order stable interpolations for immersed boundary methods. *Int. J. Numer. Meth. Fluids*, 52(11):1175–1193, 2006. Cited on page 41.

[185] X. Yang, X. Zhang, Z. Li, and G.-W. He. A smoothing technique for discrete delta functions with application to immersed boundary method in moving boundary simulations. *J. Comput. Phys.*, 228 (20):7821–7836, 2009. Cited on page 41.

[186] S.K. Doddi and P. Bagchi. Lateral migration of a capsule in a plane Poiseuille flow in a channel. *Int. J. Multiphase Flow*, 34(10):966–986, 2008. Cited on pages 41 and 145.

[187] T. Krüger, F. Varnik, and D. Raabe. Efficient and accurate simulations of deformable particles immersed in a fluid using a combined immersed boundary lattice Boltzmann finite element method. *Comput. Math. Appl.*, 61:3485–3505, 2011. Cited on pages 41, 53, 54, 55, 57, 63, 64, 65, 88, and 121.

[188] Private communication with Burkhard Dünweg. Cited on page 42.

[189] M.H. Sadd. *Elasticity: theory, applications, and numerics*. Academic Press, 2009. ISBN 978-0123744463. Cited on pages 44, 77, 146, and 147.

[190] R. Skalak, A. Tozeren, R.P. Zarda, and S. Chien. Strain Energy Function of Red Blood Cell Membranes. *Biophys. J.*, 13(3):245–264, 1973. Cited on page 45.

[191] S. Ramanujan and C. Pozrikidis. Deformation of Liquid Capsules Enclosed by Elastic Membranes in Simple Shear Flow: Large Deformations and the Effect of Fluid Viscosities. *J. Fluid Mech.*, 361: 117–143, 1998. Cited on pages 45, 47, 52, 53, 88, and 146.

[192] Y. Navot. Elastic membranes in viscous shear flow. *Phys. Fluids*, 10(8):1819–1833, 1998. Cited on pages 45, 52, and 145.

[193] D. Barthès-Biesel, A. Diaz, and E. Dhenin. Effect of constitutive laws for two-dimensional membranes on flow-induced capsule deformation. *J. Fluid Mech.*, 460:211–222, 2002. Cited on page 45.

[194] R.M. MacMeccan. *Mechanistic Effects of Erythrocytes on Platelet Deposition in Coronary Thrombosis*. PhD thesis, Georgia Institute of Technology, 2007. Cited on pages 45, 51, 61, 65, 96, 123, and 146.

[195] J. Charrier, S. Shrivastava, and R. Wu. Free and constrained inflation of elastic membranes in relation to thermoforming - non-axisymmetric problems. *J. Strain Anal. Eng.*, 24(2):55–74, 1989. Cited on pages 46 and 145.

[196] S. Shrivastava and J. Tang. Large deformation finite element analysis of non-linear viscoelastic membranes with reference to thermoforming. *J. Strain Anal. Eng.*, 28(1):31–51, 1993. Cited on pages 46 and 145.

[197] W. Helfrich. Elastic properties of lipid bilayers: theory and possible experiments. *Z. Naturforsch. C*, 28(11):693–703, 1973. Cited on page 46.

[198] R.P. Rand and A.C. Burton. Mechanical Properties of the Red Cell Membrane: I. Membrane Stiffness and Intracellular Pressure. *Biophys. J.*, 4(2):115–135, 1964. Cited on page 46.

[199] P.B. Canham. The minimum energy of bending as a possible explanation of the biconcave shape of the human red blood cell. *J. Theor. Biol.*, 26(1):61–81, 1970. Cited on page 46.

[200] R.S. Millman and G.D. Parker. *Elements of differential geometry*. Prentice Hall, 1977. Cited on page 46.

[201] G. Gompper and D.M. Kroll. Random Surface Discretizations and the Renormalization of the Bending Rigidity. *J. Phys. I*, 6(10):1305–1320, 1996. Cited on pages 46 and 47.

[202] Y. Kantor and D.R. Nelson. Crumpling transition in polymerized membranes. *Phys. Rev. Lett.*, 58 (26):2774–2777, 1987. Cited on page 46.

[203] P. Bagchi, P.C. Johnson, and A.S. Popel. Computational Fluid Dynamic Simulation of Aggregation of Deformable Cells in a Shear Flow. *J. Biomech. Eng.*, 127(7):1070–1080, 2005. Cited on pages 47, 97, and 126.

[204] U. Seifert. Configurations of fluid membranes and vesicles. *Adv. Phys.*, 46:13–137, 1997. Cited on pages 47 and 48.

[205] E.P. Newren, A.L. Fogelson, R.D. Guy, and R.M. Kirby. Unconditionally stable discretizations of the immersed boundary equations. *J. Comput. Phys.*, 222(2):702–719, 2007. Cited on page 48.

[206] C. Shu, N. Liu, and Y.T. Chew. A novel immersed boundary velocity correction-lattice Boltzmann

method and its application to simulate flow past a circular cylinder. *J. Comput. Phys.*, 226(2): 1607–1622, 2007. Cited on page 48.

[207] J. Wu and C. Shu. Implicit velocity correction-based immersed boundary-lattice Boltzmann method and its applications. *J. Comput. Phys.*, 228(6):1963–1979, 2009. Cited on page 48.

[208] X. He and L.-S. Luo. Lattice Boltzmann Model for the Incompressible Navier-Stokes Equation. *J. Stat. Phys.*, 88(3):927–944, 1997. Cited on page 48.

[209] M.E. Cates, J.-C. Desplat, P. Stansell, A.J. Wagner, K. Stratford, R. Adhikari, and I. Pagonabarraga. Physical and computational scaling issues in lattice Boltzmann simulations of binary fluid mixtures. *Philos. T. Roy. Soc. A*, 363:1917–1935, 2005. Cited on page 51.

[210] C.D. Hansen and C.R. Johnson. *The visualization handbook*. Academic Press, 2005. ISBN 012387582X. Cited on page 52.

[211] A. Diaz, N.A. Pelekasis, and D. Barthès-Biesel. Transient response of a capsule subjected to varying flow conditions: Effect of internal fluid viscosity and membrane elasticity. *Phys. Fluids*, 12(5): 948–957, 2000. Cited on page 52.

[212] E. Lac, D. Barthès-Biesel, N.A. Pelekasis, and J. Tsamopoulos. Spherical capsules in three-dimensional unbounded Stokes flows: effect of the membrane constitutive law and onset of buckling. *J. Fluid Mech.*, 516:303–334, 2004. Cited on page 52.

[213] Z.-G. Feng and E.E. Michaelides. Robust treatment of no-slip boundary condition and velocity updating for the lattice-Boltzmann simulation of particulate flows. *Comput. Fluids*, 38(2):370–381, 2009. Cited on page 53.

[214] L. Rineau and M. Yvinec. 3D Surface Mesh Generation. In CGAL Editorial Board, editor, *CGAL User and Reference Manual*. 3.4 edition, 2008. `http://www.cgal.org`. Cited on page 53.

[215] C. Geuzaine and J.-F. Remacle. Gmsh: a three-dimensional finite element mesh generator with built-in pre- and post-processing facilities, August 2009. `http://www.geuz.org/gmsh/`. Cited on page 53.

[216] E. Evans and Y.C. Fung. Improved Measurements of the Erythrocyte Geometry. *Microvasc. Res.*, 4 (4):335–347, 1972. Cited on page 53.

[217] D. Barthès-Biesel and J.M. Rallison. The Time-Dependent Deformation of a Capsule Freely Suspended in a Linear Shear Flow. *J. Fluid Mech.*, 113:251–267, 1981. Cited on page 54.

[218] J.R. Clausen, D.A. Reasor Jr., and C.K. Aidun. Parallel performance of a lattice-Boltzmann/finite element cellular blood flow solver on the IBM Blue Gene/P architecture. *Comput. Phys. Commun.*, 181:1013–1020, 2010. Cited on page 61.

[219] Private communication with Gerhard Gompper. Cited on page 63.

[220] Y. Mori and C.S. Peskin. Implicit second-order immersed boundary methods with boundary mass. *Comput. Method. Appl. M.*, 197(25-28):2049–2067, 2008. Cited on page 64.

[221] D.V. Le, J. White, J. Peraire, K.M. Lim, and B.C. Khoo. An implicit immersed boundary method for three-dimensional fluid-membrane interactions. *J. Comput. Phys.*, 228(22):8427–8445, 2009. Cited on page 64.

[222] Y. Peng and L.-S. Luo. A comparative study of immersed-boundary and interpolated bounce-back methods in LBE. *Prog. Comput. Fluid Dyn.*, 8(1–4):156–167, 2008. Cited on page 64.

[223] L. Zhu, G. He, S. Wang, L. Miller, X. Zhang, Q. You, and S. Fang. An immersed boundary method based on the lattice Boltzmann approach in three dimensions, with application. *Comput. Math. Appl.*, 2010. in press. Cited on page 64.

[224] A. Caiazzo and S. Maddu. Lattice Boltzmann boundary conditions via singular forces: Irregular expansion analysis and numerical investigations. *Comput. Math. Appl.*, 58(5):930–939, 2009. Cited on page 64.

[225] G. Le and J. Zhang. Boundary slip from the immersed boundary lattice Boltzmann models. *Phys. Rev. E*, 79(2):026701, 2009. Cited on page 65.

[226] A. Chatterjee and D.R. Heine. Numerical study of stress tensors in Poiseuille flow of suspensions. *Phys. Rev. E*, 82:021401, 2010. Cited on page 66.

[227] S.P. Meeker, R.T. Bonnecaze, and M. Cloitre. Slip and flow in pastes of soft particles: Direct observation and rheology. *J. Rheol.*, 48:1295, 2004. Cited on pages 68 and 74.

[228] P. Ballesta, R. Besseling, L. Isa, G. Petekidis, and W.C.K. Poon. Slip and Flow of Hard-Sphere Colloidal Glasses. *Phys. Rev. Lett.*, 101(25):258301, 2008. Cited on page 68.

[229] S. Gabbanelli, G. Drazer, and J. Koplik. Lattice Boltzmann method for non-Newtonian (power-law) fluids. *Phys. Rev. E*, 72(4):46312, 2005. Cited on page 73.

[230] R. Ouared and B. Chopard. Lattice Boltzmann simulations of blood flow: non-Newtonian rheology and clotting processes. *J. Stat. Phys.*, 121(1):209–221, 2005. Cited on page 73.

[231] O. Malaspinas, G. Courbebaisse, and M. Deville. Simulation of generalized Newtonian fluids with the lattice Boltzmann method. *Int. J. Mod. Phys. C*, 18(12):1939–1949, 2007. Cited on page 73.

[232] D.J. Evans and G.P. Morris. *Statistical Mechanics of Non-Equilibrium Liquids.* Academic Press, 1990. Cited on page 77.

[233] T. Krüger, F. Varnik, and D. Raabe. Particle stress in suspensions of soft objects. *Philos. T. Roy. Soc. A*, 369:2414–2421, 2011. Cited on pages 77, 79, 80, and 122.

[234] R. Skalak and S. Chien. *Handbook of Bioengineering.* McGraw-Hill, 1987. Cited on page 85.

[235] S.V. Lishchuk, I. Halliday, and C.M. Care. Shear viscosity of bulk suspensions at low Reynolds number with the three-dimensional lattice Boltzmann method. *Phys. Rev. E*, 74(1):17701, 2006. Cited on pages 88 and 96.

[236] A.Z.K. Yazdani, R.M. Kalluri, and P. Bagchi. Tank-treading and tumbling frequencies of capsules and red blood cells. *Phys. Rev. E*, 83(4):046305, 2011. Cited on page 90.

[237] I. Dierking. *Textures of Liquid Crystals.* Wiley-VCH, 2003. ISBN 3527307257. Cited on page 91.

[238] M. Tsige, M.P. Mahajan, C. Rosenblatt, and P.L. Taylor. Nematic order in nanoscopic liquid crystal droplets. *Phys. Rev. E*, 60(1):638–644, 1999. Cited on pages 91 and 92.

[239] N.J. Mottram and C. Newton. Introduction to Q-tensor Theory. *University of Strathclyde, Department of Mathematics, Research Report*, 10, 2004. Cited on page 91.

[240] T. Yamamoto, T. Suga, and N. Mori. Brownian dynamics simulation of orientational behavior, flow-induced structure, and rheological properties of a suspension of oblate spheroid particles under simple shear. *Phys. Rev. E*, 72(2):021509, 2005. Cited on pages 92 and 106.

[241] I.M. Krieger and T.J. Dougherty. A mechanism for non-Newtonian flow in suspensions of rigid particles. *Trans. Soc. Rheol.*, 3:137, 1959. Cited on page 95.

[242] G.B. Jeffery. The Motion of Ellipsoidal Particles Immersed in a Viscous Fluid. *P. Roy. Soc. Lond. A Mat.*, 102(715):161–179, 1922. Cited on pages 98 and 104.

[243] Y. Sui, Y.T. Chew, P. Roy, Y.P. Cheng, and H.T. Low. Dynamic motion of red blood cells in simple shear flow. *Phys. Fluids*, 20(11):112106, 2008. Cited on pages 101 and 145.

[244] F.M. Leslie. Continuum theory for nematic liquid crystals. *Continuum Mech. Therm.*, 4(3):167–175, 1992. Cited on page 106.

[245] K. Sakamoto, R.S. Porter, and J.F. Johnson. The viscosity of mesophases formed by cholesteryl myristate. *Mol. Cryst. Liquid Cryst.*, 8(1):443–455, 1969. Cited on page 106.

[246] B. Fathollahi and J.L. White. Polarized-light observations of flow-induced microstructures in mesophase pitch. *J. Rheol.*, 38:1591, 1994. Cited on page 106.

[247] A.D. Cato and D.D. Edie. Flow behavior of mesophase pitch. *Carbon*, 41(7):1411–1417, 2003. Cited on page 106.

[248] F. Janoschek, F. Mancini, J. Harting, and F. Toschi. Rotational behavior of red blood cells in suspension - a mesoscale simulation study. *Philos. T. Roy. Soc. A*, 2011. accepted. Cited on pages 107 and 123.

[249] E.C. Eckstein, D.G. Bailey, and A.H. Shapiro. Self-Diffusion of Particles in Shear Flow of a Suspension. *J. Fluid Mech.*, 79(01):191–208, 1977. Cited on page 107.

[250] D. Leighton and A. Acrivos. The shear-induced migration of particles in concentrated suspensions. *J. Fluid Mech.*, 181:415–439, 1987. Cited on page 107.

[251] V. Breedveld, D. van den Ende, R. Jongschaap, and J. Mellema. Shear-induced diffusion and rheology of noncolloidal suspensions: Time scales and particle displacements. *J. Chem. Phys.*, 114: 5923, 2001. Cited on page 107.

[252] J.J. Bishop, A.S. Popel, M. Intaglietta, and P.C. Johnson. Effect of aggregation and shear rate on the dispersion of red blood cells flowing in venules. *Am. J. Physiol. Heart Circ. Physiol.*, 283(5): H1985–H1996, 2002. Cited on page 108.

[253] V. Breedveld, D. Van den Ende, M. Bosscher, R.J.J. Jongschaap, and J. Mellema. Measurement of the full shear-induced self-diffusion tensor of noncolloidal suspensions. *J. Chem. Phys.*, 116:10529, 2002. Cited on page 108.

[254] J.F. Brady and J.F. Morris. Microstructure of strongly sheared suspensions and its impact on rheology and diffusion. *J. Fluid Mech.*, 348:103–139, 1997. Cited on page 111.

[255] A. Esmaeeli and G. Tryggvason. A direct numerical simulation study of the buoyant rise of bubbles at O(100) Reynolds number. *Phys. Fluids*, 17(9):093303, 2005. Cited on page 112.

[256] E.S. Asmolov. Shear-induced self-diffusion in a wall-bounded dilute suspension. *Phys. Rev. E*, 77(6): 066312, 2008. Cited on page 112.

[257] M.L. Falk and J.S. Langer. Dynamics of viscoplastic deformation in amorphous solids. *Phys. Rev. E*, 57(6):7192–7205, 1998. Cited on page 120.

[258] T. Krüger, F. Varnik, and D. Raabe. Second-order convergence of the deviatoric stress tensor in the standard Bhatnagar-Gross-Krook lattice Boltzmann method. *Phys. Rev. E*, 82:025701(R), 2010. Cited on pages 122, 139, 140, and 141.

[259] L.L. Munn and M.M. Dupin. Blood Cell Interactions and Segregation in Flow. *Ann. Biomed. Eng.*, 36(4):534–544, 2008. Cited on page 125.

[260] L.M. Crowl and A.L. Fogelson. Computational model of whole blood exhibiting lateral platelet motion induced by red blood cells. *Int. J. Numer. Meth. Biomed. Engng.*, 26:471–487, 2010. Cited on page 125.

[261] S.K. Lanka. Parallelization of the Simulation Code for a Dense Suspension of Deformable Particles in Flow via Message Passing Interface (MPI). Master's thesis, Ruhr University Bochum, 2011. Cited on page 126.

[262] S. Chapman. On the law of distribution of molecular velocities, and on the theory of viscosity and thermal conduction, in a non-uniform simple monatomic gas. *Philos. T. Roy. Soc. A*, 216:279–348, 1916. Cited on page 133.

[263] S. Chapman. On the kinetic theory of a gas. Part II. A composite monatomic gas: Diffusion, viscosity, and thermal conduction. *Philos. T. Roy. Soc. A*, 217:115–197, 1918. Cited on page 133.

[264] D. Enskog. *Kinetische Theorie der Vorgänge in mässig verdünnten Gasen.* PhD thesis, Uppsala University, 1917. Cited on page 133.

[265] M.B. Reider and J.D. Sterling. Accuracy of Discrete-Velocity BGK Models for the Simulation of the Incompressible Navier-Stokes Equations. *Comput. Fluids*, 24:459–467, 1995. Cited on pages 138 and 139.

[266] G. Hazi. Accuracy of the lattice Boltzmann method based on analytical solutions. *Phys. Rev. E*, 67 (5):56705, 2003. Cited on page 139.

[267] S.S. Chikatamarla and I.V. Karlin. Complete Galilean invariant lattice Boltzmann models. *Comput. Phys. Commun.*, 179(1-3):140–143, 2008. Cited on page 141.

[268] S. Ubertini, P. Asinari, and S. Succi. Three ways to lattice Boltzmann: A unified time-marching picture. *Phys. Rev. E*, 81(1):016311, 2010. Cited on page 141.

[269] Private communication with Gonçalo Silva. Cited on page 144.

[270] P. Bagchi and R.M. Kalluri. Dynamics of nonspherical capsules in shear flow. *Phys. Rev. E*, 80(1): 016307, 2009. Cited on page 145.

[271] K. Tsubota and S. Wada. Effect of the natural state of an elastic cellular membrane on tank-treading and tumbling motions of a single red blood cell. *Phys. Rev. E*, 81(1):011910, 2010. Cited on page 145.

[272] T. J. R. Hughes. The finite element method. Linear static and dynamic finite element analysis. 1987. Cited on page 146.

Acknowledgments

This thesis would be incomplete without mentioning a bunch of people who supported my work and listened to my professional and personal questions and laments over the past three years.

First of all I would like to thank my patient supervisors Fathollah Varnik and Prof. Ingo Steinbach who provided their scientific expertise and steadily encouraged me. Prof. Dierk Raabe, although not my official supervisor, was one of the persons initiating my project and following its progress with great interest and many suggestions. Furthermore, he arranged additional MPIE funding for my project which helped me to round off my thesis.

What would a research group be without fellow colleagues? I am indebted to all of them in so many ways. These people are: Nega Alemayehu, Segun Ayodele, Farnoush Farapour, Markus Gross, Nima Hamidi, Shiva Kumar Lanka, Suvendu Mandal, Dmitry Medvedev, Nasrollah Moradi, Mina Pouya, Alfredo Rios Nogues, and Georgios Zikos.

I would like to name several colleagues and scientists who spent a lot of time with me or gave valuable input to my work: Jörg Bernsdorf, Prof. Burkhard Dünweg, Philip Eisenlohr, Mohammad (Babak) Fallah, Thomas Franke, Prof. Jens Harting, Oliver Henrich, Dennis Hessling, Florian Janoschek, Alexandr Kuzmin, Prof. Darren Mason, Franz Roters, Prof. Matthias Schneider, Gonçalo Silva, Robert Spatschek, Denny Tjahjanto, Darius Tytko, Prof. Achim Wixforth, and Thomas Zeiser.

Furthermore, I want to thank all other colleagues in my departments at the MPIE and ICAMS, and the users of the lattice Boltzmann forum (`http://www.palabos.org/forum/`).

For a computational scientist, the IT groups are at least as important as daily coffee. Berthold Beckschäfer and Achim Kuhl (MPIE) and Niklas Caesar, Klaus Kühnberger, Vladimir Lenz, and Lothar Merl (ICAMS) provided constant support and solved my desktop, workstation, and cluster problems in the blink of an eye.

The secretaries Kirsten Berens (MPIE) and Hildegard Wawrzik (ICAMS) are the true managers of their departments. I very much appreciate their assistance in managing my official paperwork.

I also want to thank the DFG (for funding the project VA205/5-1), the MPIE (for providing me such a great base, work equipment, and additional funding), and the ICAMS (for accommodation, being my second place of work, and providing me access to its computer cluster 'Vulcan').

The most important persons, however, are Gabriele Stücken-Browers (who suggested me to apply for the open PhD position at the MPIE), my parents, Marion Sens and Heinrich Krüger, and—above all—my girlfriend Aline Browers who had to endure my daily stories about work and who constantly built me up when my motivation was suboptimal.

Printed by Publishers' Graphics LLC
BT20130108.19.20.169